# From Waterloo to Balaclava

## Tactics, Technology, and the British Army, 1815–1854

# From Waterloo To Balaclava

## Tactics, Technology, and the British Army, 1815–1854

HEW STRACHAN
Fellow of Corpus Christi College, Cambridge

*The right of the University of Cambridge to print and sell all manner of books was granted by Henry VIII in 1534. The University has printed and published continuously since 1584.*

CAMBRIDGE UNIVERSITY PRESS
Cambridge
London New York New Rochelle
Melbourne Sydney

Published by the Press Syndicate of the University of Cambridge
The Pitt Building, Trumpington Street, Cambridge CB2 1RP
33 East 57th Street, New York, NY 10022, USA
10 Stamford Road, Oakleigh, Melbourne 3166, Australia



First published 1985

Printed in Great Britain at the University Press, Cambridge

*British Library Cataloguing in Publication Data*

Strachan, Hew
From Waterloo to Balaclava: tactics, technology, and the British Army, 1815–1854.
1. Great Britain. *Army* – History – 19th century
I. Title
355'.00941 UA649

*Library of Congress cataloguing in publication data*

Strachan, Hew.
From Waterloo to Balaclava.
Bibliography: p.
1. Great Britain – History, Military – 19th century.
2. Great Britain. Army – History – 19th century.
3. Military art and science – Great Britain – History – 19th century. I. Title.
DA68.S76 1985 355'.00941 85–9843

ISBN 0 521 30439 3

CE

# Contents

# Illustrations

# Preface

When the guns fell silent on the field of Waterloo, the British army stood at a peak of professional excellence. Its own self-respect was mirrored by the admiration of Europe and by the gratitude of the nation. It was not to fight a European foe for another forty years. But when it did so, its performance was to be reviled and condemned. It had, its critics said, stood still, basking in the glory it had won under Wellington. 'The long peace' had allowed the army to atrophy in a straitjacket created by its own success.

Such an interpretation is misleading. The great powers may have been successful in keeping the peace in Europe between 1815 and 1854, but for Britain colonial commitments and the search for imperial stability necessitated constant campaigning. The British army's problems in the Crimean War did not spring simply from rust and decay.

In an earlier book, *Wellington's Legacy: The Reform of the British Army 1830–54* (Manchester University Press, 1984), I argued that in fact a considerable measure of reform was undertaken in the two decades before the Crimean War. The initial pressure for change came from two main and interrelated sources. The first was the military press. The foremost publications were the *United Service Journal*, or *Magazine* as it became (founded in 1827), whose monthly format bore a passing relationship to the famous early Victorian literary and political reviews, and two weekly newspapers, the *Naval and Military Gazette* and the *United Service Gazette* (both founded in 1833). The second was the experience of regimental officers. Following the example set by Sir John Moore at the great camp of exercise at Shorncliffe in 1803, they began to build systems of discipline on the basis of emulation and encouragement rather than that of repression and fear. Prolonged service abroad hastened the process; the regiment became a self-sufficient society, with its officers looking inwards to their men rather than outwards to the pursuits of civilian life. Most infantry battalions could reckon to spend at least fifteen out of every twenty years abroad: their officers were perforce professionalised.

Until 1852, reform tended to be specific and local. A head of steam was building up, but it was firmly contained by the aging Duke of Wellington.

If the army did rest on its Peninsular laurels, it was Wellington's fault: he denied the possibility of improving a system tested by service in war. Wellington was Commander-in-Chief of the army from 1827 to 1828, and from 1842 to 1852. Even when he was not himself in office, he retained a strong influence through Lord Hill, Commander-in-Chief from 1828 to 1842 and a former divisional commander in the Peninsula, and through Lord Fitzroy Somerset, Military Secretary at the Horse Guards (the Commander-in-Chief's office) from 1827 to 1852 and later – as Lord Raglan – Commander-in-Chief in the Crimea. Therefore, encouragement for the reformers came from other sources. Most important were senior officers, many of whom had held middle rank in the Peninsular War and several of whom had been influenced as much by Moore as by Wellington. Foremost here were Sir Charles Napier (the conqueror of Scinde) and Sir John Colborne (later Lord Seaton). In a category of his own was Lord Frederick Fitzclarence, an illegitimate son of William IV, who died in 1854 when Commander-in-Chief at Bombay. A second source was more ambivalent: the main political heads of the army were the Secretary of State for War and the Colonies and the Secretary at War. The latter presided over the War Office, and by virtue of his financial control over the army could influence much in its discipline and organisation. Enmity between the War Office and the Horse Guards was the norm, and throughout the army the War Office tended to be seen as intrusive and obstructive. Nevertheless Lord Howick, the third Earl Grey, achieved much for the army, both as Secretary at War (1834–9), and Secretary of State (1846–52).

In September 1852 Wellington died. His successor at the Horse Guards was Lord Hardinge, who had twice been Secretary at War (1828–30 and 1841–44) and who in February 1852 had been appointed Master General of the Ordnance. (The Ordnance was the third great military department: independent of both the Horse Guards and the War Office, it was responsible for the Royal Artillery and Royal Engineers, as well as guns, small arms, buildings and stores.) Abetted by Prince Albert, Hardinge threw himself into reform with an energy that belied his years. His advent coincided with a renewed fear of French invasion (the second of Richard Cobden's 'three panics', the first being in 1847–8). Parliament was therefore disposed to be liberal in expenditure, and the populace at large was supportive. Furthermore, with Sidney Herbert as Secretary at War (1852–5), peace prevailed between the Horse Guards and the War Office. Much was achieved in a short period of time: the changes that affected tactics are discussed in this book, but among the most significant of the others (at least in the present context) was the camp of exercise at Chobham in 1853 and the consequent purchase of Aldershot as a training area. The Crimean War was therefore an aberration. Strategically the diminutive British army was adapting itself to two demands – home defence and colonial wars. It did not have the resources to mount an offensive war against a European opponent; nor did it have the expectation that it would do so.

The prime concerns of *Wellington's Legacy* were the administration of the army, its structure, its recruitment and training, and its conditions of service. The present study analyses how the British army fought, and planned to fight, between Waterloo and the Crimea. The two studies are complementary, for the crucial test of any military body is its performance in combat. Therefore the argument of *From Waterloo to Balaclava* rests in part on *Wellington's Legacy*: it shows how change and development occurred in tactics as elsewhere. The only overlap in material is the discussion of Hythe and Shoeburyness (which were briefly mentioned on pages 157–8 of *Wellington's Legacy*).

Much of *From Waterloo to Balaclava* is concerned, of necessity, with the minutiae of small arms and the specifications of artillery. The introductory chapter picks up the broad strategic themes, the three-way split between defence against invasion, continental war, and campaigning in the colonies. It shows that the tension between these themes was not made explicit until 1854 itself. The succeeding chapters deal in turn with the infantry, the cavalry, and the artillery. The interaction between experience in the Napoleonic Wars, subsequent European theory (which was itself built on that experience), and imperial practice helped shape the tactics of the British army. But towards the end of the period the crucial determinant of tactical development became technological change, which effected a revolution in the power of small arms and promised another in that of artillery. These innovations were, in the parlance of economic historians, demand-driven. The army itself generated the enthusiasm for rifled firearms, although the doctrinal consequences of their introduction remained confused. Arguably, in allowing technology to determine tactical doctrine, rather than doctrine to create its own technological demands, the pre-Crimean army set the pattern for its successors up until our own day. Certainly, it was not primarily the Crimean War which initiated tactical change.

Quotations from the Royal Archives are by gracious permission of Her Majesty the Queen. Other owners of private papers to whom I am indebted include Major-General the Earl Cathcart, Dr A. E. Clark-Kennedy, R. E. Howard-Vyse, Esq., J. E. Colborne Mackrell, Esq., the Earl of Pembroke, and Lord Raglan. I must thank the Earl of Dalhousie for permission to quote from Lord Panmure's papers, now deposited in the Scottish Record Office. The Duke of Wellington allowed me to see the first Duke's papers while they were at the Royal Commission on Historical Manuscripts.

The staffs of the following archives and libraries all assisted in various important ways: the Bodleian Library, Oxford; the British Library; Cambridge University Library; the Department of Palaeography and Diplomatic, Durham University; Hereford County Record Office; the India Office Library; McGill University Library; the Ministry of Defence Whitehall Library; the National Army Museum; the National Library of Scotland; Nottingham University Library; the Public Record Office; the Royal Commission on Historical Manuscripts; the Royal Engineers Museum, Chatham; the Scottish Record Office; the Scottish United Services Museum; the East Yorkshire County Record Office. In many cases I am grateful for permission to quote from papers in their care.

I have learnt much about the infantry soldier's firearms from Dr C. H. Roads, and about the Royal Navy in general, and the Lancaster gun in particular from Dr C. I. Hamilton. Dr S. D. Badsey kindly read and commented on the chapter on cavalry, and Brigadier Shelford Bidwell performed the same service for that on the artillery. Dr Paddy Griffith introduced me to the range of related problems which confronted the French army. Finally, I must thank Clive Trebilcock, not least for bearing with some of my more esoteric enthusiasms.

HEW STRACHAN

# Abbreviations used in the notes

| | |
|---|---|
| B.M. | British Museum |
| C.R.O. | County Record Office |
| N.A.M. | National Army Museum |
| *N. & M.G.* | *Naval and Military Gazette* |
| N.L.S. | National Library of Scotland |
| P.P. | Parliamentary Papers |
| P.R.O. | Public Record Office |
| R.A. | Royal Archives |
| S.R.O. | Scottish Record Office |
| *U.S.G.* | *United Service Gazette* |
| *U.S.J.* | *United Service Journal* |
| *U.S.M.* | *United Service Magazine* |
| W.O. | War Office records in the Public Record Office |

# 1

# The theory and the conduct of war

Waterloo was won by an army that had made minimal concessions to Napoleon or to the nineteenth century. Strategically the Emperor had completely outwitted his opponents, divided their armies and left Wellington to fight a desperate, defensive battle with the Frederickian assets of well-chosen ground and skilfully deployed firepower. Napoleon had paid little attention to tactics, and it was with tactics (and Prussian good timing) that he was defeated. To subsequent British commentators, Napoleonic precepts in strategy and grand tactics were therefore overshadowed by the demonstrable defeat of column (used in 1815 in its most inept form) by line, and of cavalry by square. Furthermore revolutionary warfare, apart from being constitutionally unsettling, was blood-thirsty and profligate.

This gross over-simplification of the conduct of war was unlikely to be undermined when its serious study constituted no part of the official fabric of British military institutions. Military history assumed a very low profile at Woolwich and Sandhurst, and no other body existed to devote attention to strategic questions. The Commander-in-Chief had little direct responsibility for field operations, the Secretary at War was an accounting officer and the Secretary of State for War and the Colonies, on whom the burden of planning nominally fell, was either too concerned with colonial affairs or was hamstrung by the absence of professional advice and the lackadaisical attitude of his predecessors.

It is true that Lord Grey, Secretary of State from 1846 to 1852, did attempt to come to terms with troop dispositions on a global scale, and so began to conceive of likely scenarios for their employment. But the problem was not simply that his breadth of vision was virtually unique at this period. Each area of operations presented a radically different set of climatic and topographical conditions and the military organisation of the indigenous populations passed through a similarly bewildering series of combinations. All the parts of the British army were meant to be as ready to face Chartists as they were Kaffirs, and to cope equally well with jungle, bush and desert. Assuming the relative impossibility of training for this wide diversity of campaigning effectively, the emphasis had instead to be on ready adaptability to each new station. The important factors in colonial warfare were in consequence, broadly speaking, tactical.

British military experience between 1815 and 1854 tended therefore not so much to undermine the simplified conclusions drawn from Waterloo as reinforce them. Napoleon's grand strategic combinations rested on the infra-structure of European industrial and agricultural revolution – roads for rapid movement, root crops for supply. But outside Europe, war reverted to its eighteenth-century fundamentals. The immensely long baggage columns tied to a single route and the establishment of interim magazines were as typical of the Kabul campaign of 1841 as they were of the Seven Years War. The amount of artillery that could cope with the transport conditions was small in total and in individual weight. Manoeuvre was nonetheless circumscribed, and by the word 'strategy' a British nineteenth-century colonial general understood – just as Frederick the Great had done – little more than the organisation of marches. The remoteness of these theatres of operation from the home country meant too that, however recruited, the army became in effect a long-service professional one with a high degree of individual skill, a self-contained ethos exclusive of the broad current of national development, and an organisation in which the battalion was the central administrative, and to a lesser extent tactical, unit.

The failure to grasp the essence of Napoleonic grand tactics and strategy was not therefore an overt weakness. Divisions and corps were treated in military manuals as bodies formed for logistical convenience in large armies. The flexibility of manoeuvre they had allowed the Emperor in threatening alternative objectives, in making a mass army a series of articulated but connected limbs, was not observed.

However, although there was a proportionate tendency to play down the Napoleonic contribution, it did not preclude a conception at any rate of continental *grande guerre* orthodoxy. In Britain, as in other European countries, this was expressed through a reverence for the books of Ney's former chief of staff, A. H. Jomini. But even Jomini's impact was at best indirect. Neither the *Traité des grandes opérations militaires* nor the *Précis de l'art de la guerre* was translated into English. Lieutenant J. A. Gilbert, R.A., published a very abbreviated version of the *Traité* in 1826 but his book failed to find a market.[1]

Jomini relied instead on his English imitators to propagate his message. This meant that the principal theme of his work, a systematic analysis of his understanding of the principles of strategy, took second place to his unexceptionable but unoriginal views on tactics. Captain F. Robertson, in an article on the tactics of the three arms in that compendious and important publication, the *Aide-Mémoire to the Military Sciences*

(1846–52), never strayed far from Jomini. Lieutenant Jervis W. Jervis's *Manual of Field Operations* (1852) was more eclectic in its source material, embracing not merely Jomini but also Jacquimot de Presle's *Cours d'art et d'histoire militaire*, Lallemand's *Opérations secondaires de la guerre*, and even the experience of British colonial operations. Jervis's book, William Napier hoped, would 'turn young officers' minds to the study of the highest points of their profession and it will be a useful manual for Generals suddenly thrown into command without previous experience'.[2] But for a summary of the strategic principles enunciated by Jomini, Napier recommended Edward Yates's *Elementary Treatise on Strategy* (1852). Although apparently bought by Wellington and by his successor as Commander-in-Chief, Lord Hardinge,[3] it is nonetheless unlikely that the military fraternity took kindly to professional instruction from a mere scholar of St John's College, Cambridge.

In any case the most distinguished British disciple of Jomini was William Napier himself. Napier introduced the anglophone public to Jomini with a review of the *Traité* in the *Edinburgh Review* of July 1821, calling it 'unquestionably one of the most profound, original, and interesting that has appeared in our day'.[4] Sir William at least acknowledged the nature of Napoleon's contribution, and observed that the great advance since Frederick was in the development of strategy, not tactics. But this said, he did not feel beholden to preach his recognition to his fellow countrymen. He rebutted promptings that he should write on the philosophy of war,[5] and instead devoted his attentions to his classic, *The History of the War in the Peninsula.* Taken in conjunction with the publication of Wellington's despatches, here was ample raw material for the student of war, and the popularity of both works did much to introduce the military virtues to those outwith purely professional circles. Sir George Murray, who was pre-empted by Napier in his first ambition (that of writing an account of the Peninsular War) and instead settled for editing Marlborough's despatches, reflected a popular view. In opposing a translation of Marmont's work on the art of war in 1845, Murray pronounced that 'to beginners at least, Abstract Treatises are quite as likely to mislead the judgements, or to cramp genius, as to instruct . . . The best way in which a Military Man can improve his natural talent is by reading the most authentic accounts which can be had of the actual operations of war.'[6]

But military history, particularly on so voluminous a scale, requires intellect to unravel its precepts. William had not, unlike Henry Lloyd in his account of the Seven Years War, attempted to incorporate an analysis of warfare in more general terms. Most readers were instead provoked by the author's Whiggism and admiration for Napoleon, so much so that his brother, Charles, concluded that 'William would have done better to have described the war as one in Australia between Kangaroos'.[7]

William's failure to write in this vein is most adequately explained by his total acceptance of Jomini, whose writings in the estimation of some contemporaries rendered any further analytical work on war superfluous. But neither author attempted to rectify the dominant failure of the age. This was the lack of any attempt to reach a broad definition of strategy, its meaning and its implications, its relationship to tactics on the one hand and politics on the other. Jomini's analysis of the relation of diplomacy to war (chapter 1 of the *Précis*) was simplistic, and in any case distinguished by its being almost totally disregarded. If strategy was employed as a word (and that in itself was rare), the general intention was to indicate grand tactics, the combinations that would lead to battle, the organisation of marches and supply, the co-ordination of movements at a distance from the enemy.

Napier wrote a perceptive piece on Wellington, designed to defend the Duke from criticisms about soldiers dabbling in politics. His argument rested on the fact that

> No man can be a great captain without being at the same time an acute politician ... How is it possible, indeed, that any great military project can be undertaken with the least hope of success on any other principles? The object of all military operations must be the reducing the strength and subverting the policy of an adverse nation; and to know when and where and how to strike with effect, a general must know what constitutes the weakness and the strength of that adverse nation.

Herein were the germs of a wider definition of war, identifying the object and relating it to the means. Furthermore, Napier went on, the military system of a country will reflect its development in a broader sense; it is necessary 'to commence the reform with civil institutions, which shall bring the habits, and customs, and knowledge of the nation into harmony with their military requirements'.[8]

But Napier flattered to deceive. He did not develop these great themes. Instead he returned to his more usual subjects – the glorification of Napoleon, the reverence of Sir John Moore and the vilification of another important military writer, John Mitchell. He saw political intervention as having upset Moore's conduct of the Corunna campaign – just as Wellington blamed the politicians for the disaster that overtook the army in Afghanistan in 1841–2. He did not in the end embrace the fact that if war had any purpose it was as an instrument of policy, and that strategy at its highest level involved the integration of what was politically desirable with what was militarily feasible. The consequence of this bifurcation was the invasion of the Crimea by generals who realised that militarily they were insufficiently prepared but who, because of their lack of strategic insight, could not express their opposition effectively.

Napoleon's achievements – and arguably his failure – had sprung from his control of both war and politics, and his consequent ability to produce an integrated strategy. However, because Jomini and Napier could not elevate themselves to this plane of military thought, Napoleon's contribution was obscured from them. Of course they acknowledged Napoleon's genius for manoeuvre, but they could not divorce themselves from the contemporary tendency to work in the familiar currency of tactics, or, at best, of grand tactics within specific theatres of operations. Neither was prepared to recognise that Napoleonic warfare constituted a complete break with the past. They saw Frederick not as the perfector of the old, but as the initiator of a new system which Napoleon simply refined. On this reading, Napoleon was little more than the best of the eighteenth-century generals. The *Traité* was after all a self-avowed comparison of the two great captains, and was an attempt to rationalise in classical terms the work of a supreme pragmatist. Jomini was penning a plea for the restoration of ordered, eighteenth-century, limited warfare: 'I acknowledge that my prejudices', he wrote, 'are in favour of the good old times when the French and English Guards courteously invited each other to fire first, – as at Fontenoy, – preferring them to the frightful epoch when priests, women and children throughout Spain plotted the murder of isolated soldiers.'[9]

What emerged destroyed the Emperor's flexibility of manoeuvre. Frederick's oblique order used at Leuthen (1757) remained enthroned as the ideal order of battle. The virtues of the corps system, in forcing the enemy to disperse by threatening alternative objects and then massing with rapidity against the decisive point, became swamped by the exclusive emphasis on the latter principle alone. Division of forces was castigated, and by implication the enemy could only be turned by an extension of the line, rather than fallen on from the rear, in a classic *manoeuvre sur les derrières*. In interpreting the principle of operating on interior lines as Napoleon's preferred manoeuvre, Jomini urged that armies be kept massed but also ponderous, limited in size by the difficulty of supply, and unsubtle in their movements.

Thus, as the high priest of continental *grande guerre* orthodoxy, Jomini's impact – even if it was limited in Britain – did not undermine the conclusions of Waterloo or of colonial warfare. The pull back to the eighteenth century continued unabated. The two principal works on strategy published in the decade after the Crimean War show how deep was the trend. William Napier's son-in-law, Patrick MacDougall, the first commandant of the Staff College, based his *The Theory of War* (1856) on the campaigns of Frederick, Marlborough and Napoleon in Italy (but not subsequently). Precepts such as 'Never attack with a fraction of your force, when a short delay will enable you to attack with masses'[10] struck at the heart of the proper use of the corps system. E.B. Hamley, the first professor of military history at the Staff College, had not escaped the Jominian thrall. *The Operations of War* (1866) was a treatise for wars of limited objectives: 'The moral of this book is, not that numbers and wealth must prevail.' The object of strategy was manoeuvre, not to bring the enemy to battle. Since a general should not attack without superior forces, his opponent would not engage: 'Modern battles are for the most part partial attacks, where the assailant puts forth his foot no further than he can be sure of drawing it back again.' This was the logical conclusion to the ponderous movements to which a massed army operating on interior lines was condemned. Tied to their supply lines, Hamley saw that 'Two armies are not like two fencers in an arena, who may shift their ground to all points of the compass; but rather resemble the swordsman on a narrow plank which overhangs an abyss.'[11] Mid-nineteenth-century British military strategy, such as it was, was little more than an articulation of the problems facing an eighteenth-century army.

An early reader of Jomini in Britain was Sir John Burgoyne, Raglan's *de facto* strategic adviser in the invasion of the Crimea. His long memorandum on the *Traité* is largely a précis and it is hard to know how far Burgoyne found himself in agreement.[12] All the Jominian quirks are evident. Lloyd is praised; Guibert is criticised since his system was not – as it claimed to be – in accordance with Frederick's ideals. Strictures on combined movements from a distance and on attempting to turn a flank with part, rather than all, of an army were faithfully incorporated. But hints that Burgoyne was not in slavish agreement break out. He entered into some detail on partisans and quoted with approval the definition of the art of war as mastering the enemy's communications and then marching to battle. His principal thrust was directed on Jomini's precept that a move deep into the enemy's rear, against his communications, was ill-conceived, and that by preference a commander should concentrate his attack close in on the enemy's line. Burgoyne castigated this as 'a general maxim for what can only apply to particular occasions'. He suggested an alternative strategy which owed its origins to Wellington in Spain and to the Russians in 1812.

> Suppose an army acting on the defensive, draws an enemy well into the country, takes up strong positions that can only be attacked to a very great disadvantage & at last checks him entirely without coming to a decisive action; what can be better than to employ all the troops beyond what are necessary to check the invader in front, upon his communications in the rear?

It was a plan of this sort rather than manoeuvring on interior lines that Burgoyne proposed for the British operations in defence of Constantinople in 1854, and which he also wished to apply in the Crimea. With the sea at their backs and maintained by naval supremacy, it seemed an ideal strategy for British troops to adopt against a continental power, forcing it to sacrifice its numerical advantage by taking the offensive and its material superiority by operating at the end of a long line of communications.

Although he rarely, if ever, cited the maestro in defence of his proposals, it was nonetheless true that Burgoyne's insight had been conceived in the framework of Jomini. A total rejection of the *Traité* could give rise to problems of a different order since alternative doctrines were not so well thought out. Another Crimean general, Sir George Cathcart, was rare among British officers in that his experience was based on the 1812 and 1813 campaigns in Central Europe (about which he wrote a book) rather than on the relatively sterile fields of the Peninsula and Waterloo. He was thus aware of strategy as a distinct and important science in its own right, involving the planning of war by cabinet, council or Commander-in-Chief. Cathcart therefore acknowledged that, although Marlborough had seen war as having repercussions outside the immediate theatre of operations, the development of strategy had only been fleetingly served by Frederick. It was the large armies of the revolution that had forced preconcerted prior planning and had therefore initiated a better grasp of the fundamental principles of war. But from this unexceptionable premise, Cathcart allowed his Eastern European service to lead him to excessive admiration for Archduke Charles, the reformer of the Austrian army and author of a number of works on strategy. Because Charles had developed many of Jomini's formulae before Jomini himself, Cathcart condemned Jomini's writings as no more than a translation and paraphrase of the work of the Archduke.[13] But thus was Cathcart drawn even further into concepts of limited and defensive warfare. So great was Archduke Charles's emphasis on the safety of the base of operation, the danger of flanking or turning movements and the need for reserves, that he did not even stress that the mass should be concentrated against the decisive point.

What is strange among British admirers of continental theorists is their failure to analyse the Prussian tradition. The Frederickian army, in its atrophied state, had been crushed at Jena, and yet it was Frederick's principles to which the British still deferred. Indeed Captain Warden Flood in his *Sketch of the Military and Political State of Prussia* (1833) averred that the Prussian army was still 'esteemed the classic one of Europe' because of Frederick – not because of its subsequent reformers, Scharnhorst or Gneisenau. The reformed Prussia of Leipzig, and indeed of Waterloo, was examined in its details – light infantry tactics, military punishments and matters of that ilk – but not in its overall conception. This rejection was probably not so much ideologically motivated – although a nation in arms fighting a *guerre à outrance* would not have been sympathetic to a country enamoured of eighteenth-century military principles – as simply a reflection of an ignorance of the German language. It was not that Clausewitz was dismissed: he was simply unheard of. One of the major British works on military science mentioned him in passing only,[14] he was apparently still unknown to William Napier in 1848,[15] and a reference to him in 1852 by an officer in the 7th Fusiliers in a paper on Gustavus Adolphus can be accounted a novelty.[16]

It was as a historian rather than as a theorist that Clausewitz was appreciated. The Earl of Liverpool told Wellington in 1840 that Clausewitz's 'On War & the mode of conducting war ... [was] a work of exceeding interest', but in translating Clausewitz's volume on Waterloo Gurwood excised the 'reveries' for the Duke's benefit, as only facts were wanted.[17] In 1840, the *United Service Gazette* reported that

> Great sensation has of late years been excited in Germany by a writing 'On War' – 'Vom Kriege' – by General Clausewitz. The work might, perhaps, be called a new system of strategical philosophy, as it explains a number of strategical doctrines in a new and striking manner. The book ... is certainly one of high ability; but it is totally unknown in England, though published at Berlin as far back as the year 1832.[18]

In 1854 a translation into French by a Belgian officer rendered *Vom Kriege* more approachable to the English reader and the event was greeted by the *United Service Magazine*'s publication of Duparcq's commentaries on Clausewitz. But the fact that a full translation had to wait until 1873 shows not only British detachment from strategy *per se* but also the unacceptability of Clausewitz's emphasis on battle as strategy's central object and his belief that generals, far from embracing limited warfare as the Jominians wished, were only restrained from total war by political expediency.

Furthermore, affection for tactical rather than strategic precepts in the pre-Crimean period meant that Clausewitz's outstanding British pupil imbibed the great man's wisdom in that area alone. The redoubtable John Mitchell was a champion of the German school rather than the French. The works of Marbot, Jomini and Rogniat demonstrated, he wrote, 'that men of ability may be present in twenty campaigns without comprehending the first principles of their profession'.[19] Mitchell instead described himself as 'a humble follower'[20] of Behrenhorst, author of *Betrachtungen über die Kriegskunst* (1798–9) and an officer who, in recognising war's uncertainties and placing these at the centre of his analysis, was the precursor of Clausewitz himself. Mitchell's own admiration of Clausewitz, like that of most of his readers, grew with his acquaintance and comprehension. In 1838, he called him 'a general officer of the highest talents' who had written 'a very able, though lengthy, and often obscure book on War'.[21] By 1842, he avowed that the work of Clausewitz formed 'an era in military science'.[22] Although he ignored a consequent appeal that he translate *Vom Kriege*, he continued to trumpet Clausewitz's contribution as 'the highest authority that can be quoted on a military subject'.[23]

Mitchell might bemoan that 'no one [in Britain] has written a word on what may be termed the speculative part of the profession',[24] but his own *Thoughts on Tactics* (1838) suggests that he too had but limited qualifications. Jomini was guilty of a misinterpretation of Napoleon, but had at least acknowledged his pre-eminence. For Mitchell, Frederick alone was the skilled strategist. Revolutionary warfare's 'only discernible principle – the only principle from which it never deviated, was an utter disregard of human life and human suffering'. It mistook 'the effects produced by the force of arms and of numbers for the effects of skill and combination'.[25]

The failure to recognise Napoleon's supremacy as a strategist meant that Mitchell's own comprehension of strategy's ramifications was primitive. At the level of grand tactics, his scornful pen cast a veil over the corps system. 'Even the greatest of the Continental battles', he wrote,

> lasted entire days; they were fought for the possession of posts or villages on which the world's fate seemed to depend, or by swarms of skirmishers only, while armies remained in reserve, literally to feed the fire – to come, like the wounded Curiatii, successively, instead of simultaneously, into action; and above all, to save the fame of their mighty leader from too severe a shock whenever the goddess of fortune remained deaf to the progress of her votaries, though uttered in the dying groans of idly-sacrificed thousands: one bold onset would have been worth all this strategy a hundred times over.[26]

So, for Mitchell, the joining of battle with all available resources as soon as possible was the key to strategy. In fairness to him, he was not alone in this simplistic interpretation of Clausewitz: Ludendorff represented the consummation of a Prussian tradition that the total commitment to battle alone was the centrepiece.

For, like the German general staff, Mitchell did not understand Clausewitz's observations on what is now called grand strategy. Clausewitz felt that the horror of the cauldron thus brought to boil was in practice mitigated by political objectives, by the art of the possible. But for Mitchell, political considerations in military matters were a tiresome intrusion. He, in common with other British soldiers, opposed the army's practical (if not theoretical) dependence on Parliament, and the consequently fluctuating dictates of party policy in military affairs. Mitchell bitterly resented the military's subordination to politics, failed to see the relationship of war and politics, and in his anger failed to put politics at the centre of his analysis.[27]

Instead he started from the bottom up. Since the purpose of strategy was the decisive engagement, and its definition the theory of the use of combats for the object of the war, it followed – in Clausewitz's view – that tactical developments would have their repercussions on strategy. This was a healthy corrective to Jomini's search for general operational principles, but Mitchell was just as capable of uncluttered assertion. He distorted Clausewitz sufficiently to conclude that tactics constituted 'the legs and arms of the whole science of war'.[28] Battle having been joined as rapidly as possible, the strategic aim was subordinated to the tactical means in the fulfilment of victory. Thus, however vituperative his abuse of contemporary British military thought, Mitchell had worked himself into a position that was concerned with the same aspects as Napier and his imitators.

His contribution, derived from his reading of Behrenhorst and Clausewitz, was to examine tactics not as a systematiser attempting to impose order (as had British drill masters, like Dundas, in the 1780s and 1790s), but as a child of the age of revolution taking the individual as the prime unit. 'If, as the poet says, "the proper study of mankind is man", to no class of man is that study so essential as to all ranks of officers; for a knowledge of war is, in fact, very little more than a knowledge of what man can be brought to perform.'[29] His views on infantry training reflected the central position accorded to the individual's capabilities and initiative by the appearance of light infantry on the battlefield. He condemned Dundas's adaptation of Frederickian drill to Britain in terms which would have done credit to Yorck, Scharnhorst and others of the Prussian reform movement. 'Modern tacticians', he complained,

> dream of nothing but mechanical precision; and totally forget that the automaton machines, whom they think of moving by mere rule and square, have each and all lives, souls and feelings implanted in them by the hand of their Maker, and not exactly to be eradicated at the simple command of adjutants, adjutant-generals, or drill-sergeants. This is no doubt a great evil; but as we cannot change the works of providence, we must just try to adapt our poor and puny science to its omnipotent decrees; and, instead of attempting to extinguish in living men, all human character and feeling – call it feebleness if you like – we must seek to regulate our system according to that character. Psychology and a knowledge of the effect of arms, now totally left out of sight, thus become the basis of a just system of tactics.[30]

In this particular, Mitchell breathed the pure spirit of Clausewitz. Moreover, what he said accorded with the legacy of Sir John Moore. Moore, the hero of many thinking officers in the British army, had recognised the need to nurture rather than crush individuality in training, and had imparted this in the camp of exercise at Shorncliffe in 1803. But whereas most of

Mitchell's contemporaries continued Moore's work by endeavouring to improve the soldier's conditions of service, Mitchell himself applied Moore's precepts to tactical thought.

The nature of his insight, and its relationship to revolutionary warfare, were not grasped by Mitchell himself. For him, Clausewitz was not the spokesman of a group of Prussian reformers trying to replace Frederick's legacy with the tactical ideas of the *levée en masse*, but an heir, through Behrenhorst and his experience in the Seven Years War, of one branch of the Potsdam tradition. His examples and authorities were therefore derived from the eighteenth century. Marshal de Saxe, admittedly a far-sighted soldier, but whose *Reveries upon the Art of War* was published in 1757, was as fruitful a source for Mitchell as for numerous of his contemporaries. Folard, whose treatise on Polybius had first put the case for shock rather than firepower in its espousal of the massed infantry column, was also praised. But Guibert's *Essai général de tactique* (1772) and Bourcet's memoranda (1764), which are considered to have exercised a profound influence on Napoleon's thought, were not even mentioned. Bourcet's work had been published in the *Journal des sciences militaires* and was used by Captain W. C. Mayne in his book, *Military Reconnaissance* (1837). It was therefore known in Britain, but its early advocacy of a divisional system for use in Alpine passes can have struck few chords in Mitchell. Guibert, although ostensibly rooted in an admiration of Frederick, in fact represented a break with the Prussian influence. He was rather the consummation of a French tactical school, ruminating since it had been vanquished by Marlborough. Mitchell's omission of his work was especially significant, for to Guibert were traced the origins of the 1791 French ordinance on drill; in the dialectic between the shock of massed column and the firepower of the thin line, this struck a balance marginally in favour of the latter and had advocated shallow columns. It had thus provided the framework for revolutionary warfare. In consequence, Mitchell's theses, though based on a realistic appreciation of the capabilities of smooth-bore firearms, emphasised shock over firepower, and could not outlive the imminent advent of the breech-loading rifle to the battlefield. This tendency was most manifest in his arguments for the capabilities of cavalry *vis-à-vis* their pedestrian rivals. As did Captain L. E. Nolan in *Cavalry: Its History and Tactics* (1853), Mitchell found the most fruitful buttress for his arguments were the achievements of Frederick's horsemen, 'unsurpassed examples of cavalry excellence', in direct contrast to 'those feeble modes of combat' of Napoleon's cuirassiers.[31]

Mitchell therefore conspired in the national throwback to Frederickian tactics which Wellington and Waterloo had encouraged. The choice of his eighteenth-century authorities was not very different from that of his arch-rival, William Napier. In fairness, the latter at least acknowledged Guibert, but he did not place him on the pinnacle of eighteenth-century military thought which he now occupies. Both respected Lloyd,[32] the Welshman who had reached general's rank in Russian service, and had then written an account of the 1757–9 campaigns and his extrapolations therefrom, entitled *History of the Late War in Germany*. Lloyd, Napier wrote, 'has not only treated [the art of war] as a philosopher, but was the first, perhaps, to show that war, however uncertain in its issues, had principles without which no great success could be achieved'.[33] The other historians of Frederick's campaigns, Warnery and especially Tempelhoff, were cited alongside Lloyd, Behrenhorst and Bulow, and pre-Crimean military thought therefore reflected the experience and issues to which those men attached importance, those of eighteenth-century tactics.

William's brother, Charles, provided at Miani[34] in 1843 an excellent example of this inheritance functioning in practice, and, by leaving the field victorious, despite desperate odds, showed too its compatibility with the demands of colonial warfare. For him, much more than his francophile brother, Frederick was the model, and it was to Seydlitz, the Prussian cavalry leader, not Murat, that he likened his own commander of horse, John Jacob. The Baluchis, in their well-chosen position in a dried-up river bed, with flanks resting on strong natural features, acted in a fashion reminiscent of an eighteenth-century opponent. Napier reflected the challenge in his manoeuvres. The security of the Baluchi flanks seemed to make the centre the only avenue for attack, but in any case Napier might have discountenanced a flank attack. 'Turning a flank', he wrote, 'is of no use, except to force an enemy to quit a strong position and occupy a weaker.' He admitted that a flank attack might enable a general to cut an enemy's communications, but the Baluchis had none. Napier's purpose in adopting that distinctive Frederickian manoeuvre, the echelon attack, was therefore not to gain a flank or deceive the enemy, but to protect the refused left wing from the threat of the village on its flank, while bringing his best troops, placed at the advance of the echelon, into action as quickly as possible. Thus, while containing his flanks, his strength could be directed against the weaker centre and the Baluchi wings thereby divided. The only attempt to obscure these movements was to cover the infantry's deployment by artillery fire – yet another eighteenth-century feature.

Napier envisaged carving his central hole not by firepower but by shock, using the superior morale of his troops in conjunction with the bayonet. 'The short range and very uncertain flight of shot from the musket', he wrote later in life, 'begets the necessity of closing with the enemy, which the British soldier's confidence in superior bodily strength, due to climate, pushes him to do.' He felt in consequence that 'No troops in the world will stand the assault of British troops, if made with the bayonet and without firing.'

In fact, on discovering the mass of Baluchis in the bed of the river Fullaillee, the infantry withdrew to the edge of the bank and poured in successive volleys, aided by the guns which were able to enfilade the position. Napier's entreaties to his men to close were ignored. Major Waddington, one of the participants, concluded that 'the bayonet was but little used except in defence, and it shortly became evident that the fire of the matchlock and the glancing of the keen sabre were less and less frequent, while the continued and destructive roll of musketry, delivered from the edge nearly of the river bank, levelled every living being before it'. The decisive stroke was then delivered. Ironically, it came not from the central infantry assault but from a cavalry charge on the Baluchis' right flank. This completed the dislocation already wrought by the fire of the infantry, who were thus able to put in their final attack. Vital as the psychological determination of Napier was, the real elements of his success were firepower followed by a flank attack to weaken the centre.

Nor was Miani an isolated example. Gough, who commanded the armies in the Gwalior and Sikh Wars, had little strategic sense, no capacity for administration and spurned the development of firepower. But his determination to advance was carried through by the high morale and well-drilled tactical proficiency of his men. At Maharajpore (1843) a frontal assault against a strong position with minimal artillery preparation and when a flank attack was possible set a pattern that was to be repeated up until Gujerat (1849). At Mudki (1845), Gough advanced directly on an unknown number of Sikhs already in position. At Sobraon (1846), Gough again spurned a turning movement for an attack all along the line. What at one level shows strategic daring, at another demonstrates unbridled folly. The invasion of Gwalior in 1843 was conducted by two separate forces, converging from different directions. Such a departure from Jominian orthodoxy worried Harry Smith, a great proponent of mass on the decisive point, but he concluded that it probably did not matter, 'for strategy is totally unknown to a native army, which usually posts itself in a well-chosen position and awaits an attack'.[35]

Eighteenth-century tactics were thus adapted to and modified by colonial warfare. Herein lies a central debate in the military thought of the pre-Crimean army. Furthermore colonial fighting could in some cases result in British thinking being in advance of European. The French between 1815 and 1851 made little attempt to analyse Napoleonic experience, but rather reverted to the tactical arguments of the *ancien régime*, looking to Frederick and Guibert.[36] The attributes the British admired in the French army in the Crimea, and in particular the French soldier's resourcefulness and self-sufficiency, were not those gleaned from *grande guerre*, but those distilled from Algerian experience. Similarly the British army represented a blend of Wellington's version of eighteenth-century thought with a whole variety of colonial campaigning.

This theme is well illustrated by the spate of books on light infantry drill, outpost work and field fortification. In immediate terms the writings reflected the interests of the best schools in the Peninsula (the 43rd and 52nd Light Infantry, and the Rifle Brigade, of the Light Division), but at one stage removed they were the interests of Sir John Moore and the Prussian reformers, the acquirements of individual training and the topics on which the regulations were thin. These too were the areas which the bush-fighting of colonial warfare often helped to develop. Thus was the army pulled in one direction by the formalism of Frederickian drill, the *grande guerre* orthodoxy, and in the other by the experience of campaigning against native enemies. The continental school could accept as confirmation the fights against well-organised armies like the Sikhs, but might reject much that was gleaned in South Africa or New Zealand as an aberration in European terms. At least, however, colonial campaigning ensured that the skills of skirmishing and the employment of mounted infantry were tested in advance of the supplantation of the musket by the rifle, which thus brought the victory of firepower over shock.

The debates discussed in the ensuing chapters reflect these concerns. The tactical problems encountered by colonial generals and their relationship to Peninsular experience were the nub of the dialectic in military thought. Furthermore the argument did not necessarily take the form of a rigid division between arch-Europeans, rejecting any experience gleaned elsewhere, and youthful radicals convinced that skirmishes in the bush could be equated with great land battles. The fact that so many middle-ranking officers of the Napoleonic Wars reached the summits of their careers in the colonies meant that the selection of relevant experience was a personal difficulty as well as a public debate. Lord Hardinge and Sir George Cathcart were both officers who had served with continental armies in 1813–15 but who nevertheless based much of their thinking on what they saw in India or at the Cape. By contrast, however, grand tactics and strategy were of little interest or importance. At the former level, a few books discussed divisional or brigade drill, but in 1854, in default of any other guidance for the staff, Hardinge had to order a reprint of Murray's Peninsular instructions to officers of the Quarter Master General's department.

On the higher plane, there was simply no indigenous school of strategy. The *Aide-Mémoire to the Military Sciences* tried to articulate the idea of a specifically British form of warfare, an adaptation of 'general principles completely in harmony with the general maxims'. It postulated that

> A British military writer may view the questions involved in the term 'great operations' ('grande tactique' of the French) either as they are based on the general principles of the science, in the light they are viewed by continental strategists, or, narrowing the subject, take it up on the insular position of the empire and the local conditions which resulted from it. For the one leads to operations of immense armies and objects which menace the very existence of states, while the other contemplates principally defensive measures at home, assistance to an ally abroad, and offensive expeditions to distant countries; mainly depending on the superiority of the Royal Navy, and with land forces in no case amounting to more than 50,000 national troops.[37]

But this was a fleeting insight in a book written primarily for sappers, the soldiers least likely to hold independent command, and which itself contained a condemnation of the amphibious means which such a statement of objectives in British warfare implied.[38] The Crimean War, since its centrepiece was an attack on the Russian naval installations at Sebastopol in order to ensure allied naval hegemony in the Black Sea, was a perfect example of what was later to be called 'the British way in warfare'. But to an understanding of this level of grand strategy, no British soldier yet aspired. Had it been seen, the inevitability of amphibious operations against an unreconnoitred, hostile and fortified peninsula would have prompted a clearer statement of the universal opposition to the Sebastopol operation. All the senior officers with the expeditionary army were averse to the plan, which was formulated by the navy, but their lack of cogency and the in-built institutional deficiencies meant that the point was insufficiently made. No clearer testimony to the bankruptcy of strategic thought in Britain needs be provided than the total inadequacy of the means applied to the objective of the Crimean War.[39]

# 2

# Infantry: the rulers of the field

In 1815, the reputation that the British foot-soldier had won in the Napoleonic Wars placed him firmly at the pinnacle of professional estimation not only in his own country but also in the rest of Europe. The qualities that had made the Peninsular infantry, according to William Napier, were not superior tactics or better armament but endurance and discipline, attributes rooted in the national character. Jomini came to similar conclusions.[1] Nonetheless, it was all too easy, in seeking to explain the British superiority, to neglect morale and look instead to quantifiable factors. Thus, in 1827, Colonel Juchereau St Denys urged the French Ministère de la Guerre to adopt the British infantry's tactical manual, the 1824 *Field Exercise*, allegedly the apotheosis of Peninsular experience in its embodiment of celerity and precision.

British views of the 1824 *Field Exercise* were less complimentary. It was designed to consolidate, not innovate. Its author, Sir Henry Torrens, never claimed that it was more than a revision of the first *Field Exercise*, that of Dundas, published in 1792. During the war, regiments had evolved their own 'improvements' on Dundas, and Torrens's purpose was to incorporate these so that the practice of drill would once again be uniform. His main provision was for a line of two, not three ranks, but there were few further changes.

In 1833, the *Field Exercise* was revised once again. Although the repetitious movements and long words of command preserved by Torrens were reduced, the 1833 *Exercise* could not be described as revolutionary. Dundas's principles had not been lost sight of, and there remained initial errors uncorrected. Even the French, though still complimentary, found the manoeuvres 'un peu compliquées pour le Champ de Bataille'.[2] The general principles and the theory of drill, which had been excised from Dundas's work on its being officially adopted, still found no place. In addition the book considered only the drill of a separate and isolated battalion, and neglected larger formations. To all these besetting sins was added shoddy presentation: the revising board had not been able to see the work through proof stage, with the result that sixty-two of its pages bore mistakes or inconsistencies.[3]

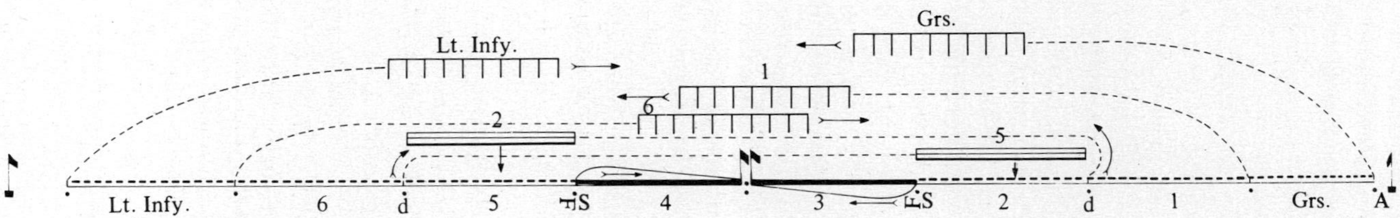


Fig. 1 *A Line Changing Front to the Rear on the Centre*: approved method without 'clubbing'. From *Field Exercise and Evolutions of the Army* (1833)

In September 1852, Lord Hardinge succeeded the Duke of Wellington as Commander-in-Chief. Revision of the infantry *Field Exercise* was one of a number of items on a lengthy agenda of reform. In 1853, a divisional camp of exercise was held at Chobham, and Hardinge told its commander, Lord Seaton, that he should use the opportunity to test the various improvements proposed for the infantry's evolutions.[4] Although it is not known along what lines official views were developing, modifications rather than major changes seem to have been envisaged. Nonetheless, they were sufficient to play their part in provoking the resignation of the reactionary Adjutant General, Sir George Brown, in January 1854.[5]

Whatever the manual's faults, it was the only tactical work that was universal and compulsory in its application. The detached nature of the British army's service meant that the drill book had to be uniformly and rigidly enforced so that when battalions came together for the first time they could manoeuvre with comfort. The proper place and time to learn this 'mechanical portion, or ABC', a reforming general, Lord Frederick Fitzclarence cautioned, was 'in time of peace on the level surface of the parade ground, with a view of applying such knowledge in broken undulated ground'.[6] But Fitzclarence's eminently practical view provided justifications for the excesses of martinets. As one military writer, Colonel Firebrace, cynically remarked, 'The first thing you do when you catch a peasant, is to make him look as if he had swallowed his grandmother's poker.'[7] Thus even the useless manoeuvres were said to be of value since their practice, although they were never performed in battle, imparted flexibility and instant obedience.

Furthermore, the field movements remained so slow and convoluted that in battle corps often lacked the time to perform the manoeuvre required. In particular the front and rear ranks were not interchangeable, and the companies had always to be numbered from right to left. Officially, therefore, if enemy cavalry attacked the rear of a battalion, a square could not be formed on the nearest companies of the rear rank. At Chillianwallah in the Second Sikh War, when the 61st was threatened from behind, its brigadier, Colin Campbell, correctly ordered the front rank to the rear, but the regiment's commanding officer, realising the imminence of the danger, simply ordered right about face, so that the rear rank became the front rank.[8] In such dangerous circumstances, this so-called 'clubbing' of the battalion (that is inverting the order of the companies) was permissible. But the central argument of some officers, and in particular Captain W. J. B. Mountsteven, was that recourse to expedients proved the uselessness of the regulations. When the troops received a command that would thus confuse their order, they hesitated and desperation replaced coolness. A corps should 'always move by the

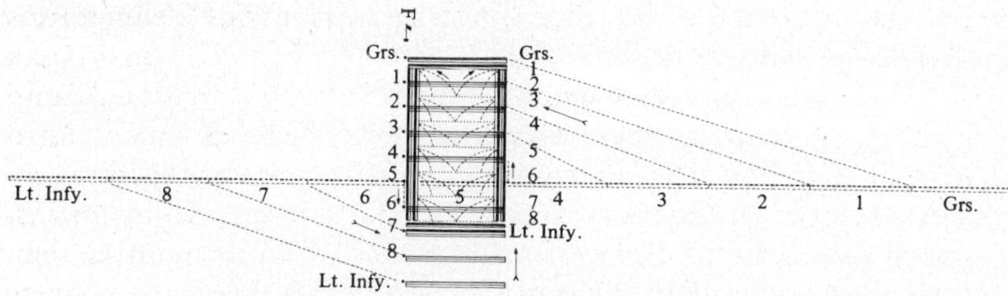


Fig. 2 *A Battalion from Line Forming Square on a Central Company.* From *Field Exercise and Evolutions of the Army* (1833)

shortest and most direct way from the formation it may happen to be in, without any regard to its being what is called "clubbed", or as to which rank is in front'.[9] Mountsteven's tenets had distinguished support. They were embodied in the *Aide-Mémoire to the Military Sciences.* Sir Hussey Vivian had urged the committee responsible for the 1833 revision to make provision for inversion against sudden attacks from different fronts,[10] and on his inspections he would require regiments to perform these non-regulation manoeuvres.[11]

In the demand for simplification there were two strands. The logical conclusion to the Mountsteven argument was that all orders should be geared not to the movements of a battalion but to those of the individual, as modern drill is. Parallel, but not as drastic in its implications, was the gradual process of cutting away the dross in the *Field Exercise*, as had been attempted in 1824 and 1833. Sir Charles Napier, to whom Mountsteven's book was dedicated, said that 'an advance in line – column – echellon (direct and oblique) and a ready formation into line at once during any movement are all a corps requires for action!'.[12]

Such reservations about the 1833 *Field Exercise* suggest that, although unrevised by 1854, drill had in practice undergone a process of gradual evolution. A French officer at Chobham reported that the English 'vont plus directement au but dans les manoeuvres et ont moins de mouvements préparatoires',[13] an opinion that few forward-thinking British officers would have endorsed but which perhaps reflects some sort of improvement. The old school observed that movements had become 'unnecessarily lax and disjointed'. Colonel W. Cochrane commented in 1852 that 'Commg. Officers of the present day seem to think that if they can only place their Regt. in the position required it matters not how slovenly the execution is performed'.[14]

The immediate explanation for Cochrane's opinion is the instruction in the 1824 *Field Exercise* that, when line battalions were perfect in line manoeuvres, they should practise 'in certain extended formations': line infantry should therefore be ready to act as light infantry. Clearly the ponderous, formalised drill of Dundas was inapplicable in broken country, in woods or amongst mountains, for the protection of outposts, or for advance and rear guards. Prompted by experiences in America since the Seven Years War and systematised by Moore at Shorncliffe, the widespread use of light troops in the British army had been the outstanding experience gained in the Peninsula. Quite apart from the constant picquets, even in a set battle Wellington had deployed an average of 1,200 to 1,500 skirmishers to screen a front of 5,000 infantry.[15] To meet these demands, the army emerged in 1815 with only six specialist light infantry regiments, the 43rd, 51st, 52nd, 68th, 71st and 85th, although the 90th was added in that year and the 13th in 1822. All of these were armed with smoothbore muskets, similar to those of the rest of the infantry. In addition there were four battalions equipped with rifles, for precision shooting, two in the 60th Rifles and two in the Rifle Brigade. As some compensation for these deficiencies, one of each regiment's two flank companies was formed of light infantry. However, to belong to it or to the grenadier company (the other flank company, which had ceased to have a separate tactical role) was an accolade that deprived the rest of the regiment of the best men. Furthermore it was probable that, given the scattered nature of the army's imperial obligations, the nearest true light infantry regiment would be many hundreds of miles away, and a single company was unlikely to be sufficient to meet the requirements of colonial warfare against a native enemy, when the whole battlefield might constitute a series of unrelated skirmishes.

The 1824 instruction therefore was not merely an embodiment of Peninsular experience but also a provision for the exigencies of colonial warfare within a reduced army. In 1834 the Horse Guards reminded the line that it would in future be required to perform light infantry manoeuvres at the half-yearly inspections, and the initial reaction of one observer on seeing the 1833 drill performed was to note its similarity to the light infantry drill.[16] Certain officers completely absorbed these instructions. Lord Frederick Fitzclarence drilled his men so that the nearest troops always pushed ahead to skirmish, thus saving the energy of those further away and in the process of realigning.[17] Sir Charles Chichester was so engrossed in the light drill of his own regiment, the 81st, that at a field day he employed it rather than the 68th as skirmishers; the 68th was a properly constituted light infantry regiment, and proved highly incensed at being passed over.[18] The flexibility these officers gave the army was invaluable: at Multan the 32nd extended three of its companies in skirmishing order,[19] at Gujerat the 24th was enabled to skirmish with its number seven company as well as with the light company,[20] and even more important, in

Fig. 3 *Attack of the Kaffirs on the Troops under the Command of Lt. Col. Fordyce of the 74th Highlanders, while Forcing their Way through the Kroomie Forest on the 8th of September 1851*: a graphic depiction of the difficulties of bush war. Drawn by Henry Martens after a sketch by Thomas Baines, and engraved by J. Harris (National Army Museum)

the desultory, guerrilla warfare against the Kaffirs, the 74th could make the best use of 'everything that afforded shelter', moving from cover to cover, and firing as opportunity offered.[21]

However, if it was available, there was a natural preference for employing a proper light infantry regiment, and in particular a rifle regiment. The Kaffirs lived in dread of the Rifle Brigade. At the Alma, it was the Rifles and the light companies who went ahead to skirmish. Such discrimination meant that many regiments neglected to practise the light infantry drill at all. The vagueness of the 1824 regulations encouraged this laxness. There was no incentive to resort to one of the plethora of unofficial manuals on light infantry and outpost work when their systems were so different that it was said no two regiments were alike. Furthermore many argued that it was pointless to train as light infantry when, without rifles, they could never be anything other than a bastardised version of the ideal *tirailleurs*. Indeed the light infantry regiments, themselves without rifles, did on occasion claim that they were trained too much as line infantry and not enough as marksmen.

The closely defined role of light infantry was derived from European precepts. The Duke of Cambridge, who was brought up in the military traditions of Hanover, as late as 1851 brigaded all the flank companies in the Dublin district to give him a light infantry battalion, a practice reminiscent of the 1770s.[22] The *grande guerre* orthodoxy, that line infantry could repel cavalry but light infantry could not, was nonsense in the context of the British juxtaposition of roles. Lopez Suasso Diaz de Fonseca, in his monumental *Theory of Infantry Movements* (1846), only mentioned light infantry once, and then to assert that they should be distinct from the main force. Major Troubridge, who was to have both legs shattered at Inkerman, wrote in his edition of a French tactical work, Lallemand's *Principles of the Minor Operations of War*, 'The fact must never be lost sight of, that bodies of skirmishers, although intended to prepare and facilitate the means of dealing the principal blow, are not themselves to deal it.'[23] By contrast the Cape taught Sir Harry Smith the value of light infantry in colonial warfare, and he concluded, in direct opposition to the continentals, that they should be trained 'for cover or for assault'.[24]

The need for infantry trained as skirmishers was admitted by both parties, but those whose ground rules were based on the Peninsula and Europe tended to conclude that the solution was more light infantry and rifle regiments, still separate from the line infantry. To fire accurately, to act independently, and to move quickly seemed to some to be natural attributes rather than qualities imparted by training: to demand that all infantry be fit to act as light troops 'seems very like expecting that a dray horse and a racer would run well in a curricle, or compete in a race'.[25]

This logic imposed itself on Wellington's Quarter Master General in the Peninsula, Sir George Murray, who saw the 'several Regiments which have the appellation of Light Infantry ... [differing] scarcely in anything but in Name from the other Infantry Regiments of the Line'. What he wanted in order to check the threat of French invasion in 1845 was a large number of men, ready to exploit the undulating and hedged English countryside, and 'specially prepared to Light Troop service'.[26] Wellington too harped back to eighteenth-century practice. He agreed with Murray on the need for genuine light infantry, but proposed that one rifle company be added to every three battalions, so that when brigaded they would have a light battalion of one rifle and three light companies.[27]

These two distinguished warriors failed to conclude that the similarity between light infantry and the line might not be so much a criticism of the former as an indication of the extent to which the latter had revised their movements. Murray himself noted that Sir John Moore had told him that 'his wish [at Shorncliffe] had not been so much to form regiments of

Infantry differing from those of the rest of the Army, as to introduce into the British Infantry in general a system more suitable than the unnecessarily slow and formal system which Sir David Dundas had copied from the Germans'. The *United Service Journal* in 1831 urged all regiments to train exclusively as light infantry, so as to conform to the realities of colonial warfare.[28] If skirmishers were to act with the line and vice versa, the argument that training infantry in both duties was either beyond their capabilities or would make them too loose in line was not applicable. As Captain Thomas Ramsay concluded in 1842, 'It is quite plain that there is only one sort of infantry, and in the modern system of warfare every foot soldier is required to perform every sort of military service'.[29]

The practice of many regiments implemented the theory. In 1852, Lieutenant J.-W. Jervis pointed out that 'in the British service, both infantry of the line and light infantry receive an identical instruction, are obliged to perform the same duty, and are armed in the same manner'.[30] In the year that he wrote, the adoption of the Minié rifle throughout the infantry finally made it clear that the distinction of light infantry was mere form. The small size of her army had already forced Britain's battalions into a practical application of this rule, but it was a technological innovation that eventually caused Wellington to accept that all infantry had become riflemen. The new firepower meant that against a foe similarly armed extended order would be the foot soldier's *métier*.

This essential truth, although immediately grasped by some, was only slowly incorporated into British infantry tactics. France remained the likeliest European enemy, and against her columns of infantry the Wellingtonian application of the line had the merit of proven worth, particularly in its development of firepower. Its success rested on the utter solidity and coolness of the individuals who formed it; its undemonstrative firmness seemed as indicative of, and as suited to, the national character as did the charging massed columns reflect French excitability. Just as light infantry drill made the line's application looser and consequently, to some eyes, more vulnerable, so even more did extended order seem anathema. 'It must be ever remembered, that steadiness in the individuals, and compactness and solidity in the body collected, are the great criterion of infantry; and that those prominent qualifications must suffer in proportion as the troops are more rapidly moved.'[31] To protect this solidity, the Duke insisted that the Minié be called a 'rifle musket', so that any idea that the ordinary infantry had become riflemen might be checked.

The case for solidity was reinforced by the British use of the line. Although Dundas stipulated a line of three ranks, in the Peninsula two had been the norm, thus enabling the additional men to lengthen the front, and add to the chances of outflanking or outshooting the enemy. Two ranks were thus a national hallmark, but pride did not lead to rashness and squares against cavalry were formed in four ranks. At the battle of Balaclava the 93rd, the 'Thin Red Line', was in fact four deep.[32] This use of four ranks against cavalry was simply to impart solidity, and even the role of the third rank, in the British view, was not to increase the volume of fire but only to prevent the front two breaking.

Quite apart from its more extended front, the British argument held that two ranks made for more efficient shooting. Although three ranks would release more rounds in a given area, especially in the vital first volley, the barrels of the third rank were sufficiently short not only to render their bayonets ineffective but also for the blast from their muzzles to endanger the front rank, particularly in the course of sustained firing. The tendency to fire high was therefore encouraged and ammunition consequently wasted. The French practice was for the rear rank to load for the second rank. Trials in Britain in 1839 showed that this only increased the fire's effect by one-eighth on that of two ranks alone, that even that was not long maintained, and that it was productive of greater inaccuracy.[33] The alternative, as happened in the square, was for the front rank to kneel, but 'this practice is excessively dangerous in the presence of an enemy, and possesses the additional drawback of interrupting the continuation of the firing while the arms are being reloaded'.[34]

However, in addition to its apparent frailty, the two-rank line suffered another and more grievous defect. It was extremely inflexible: an advance over broken ground would soon open up gaps and, if penetrated, it was hard to reform. John Mitchell had observed the sense of isolation the soldier experienced if his neighbour fell. As an immediate solution, he advocated proper training to develop his initiative, make him nimbler and more dextrous in the management of arms. More far-seeingly he argued that by extending the line the individual target was reduced and each man had room to load and fire without jostling the man next to him. Mitchell followed de Saxe in favouring the Roman system of maniples, and interestingly Sir George Murray reflected him a decade later, when he praised the use of divisions, likening them to the Roman legions, and seeing the existing 'German' drill as inappropriate for them. 'Each Division forms ... a section of an Army complete within itself; and an aptitude for prompt movements is thereby afforded which no other organization of a great army is capable of producing.' By formalising the gaps that might in any case occur in a line, Mitchell and Murray were proposing a formation that retained all the firepower of the line but at the same time had the resilience to resist cavalry. If one division were smashed, it would not be a fatal blow to the rest of the formation.[35]

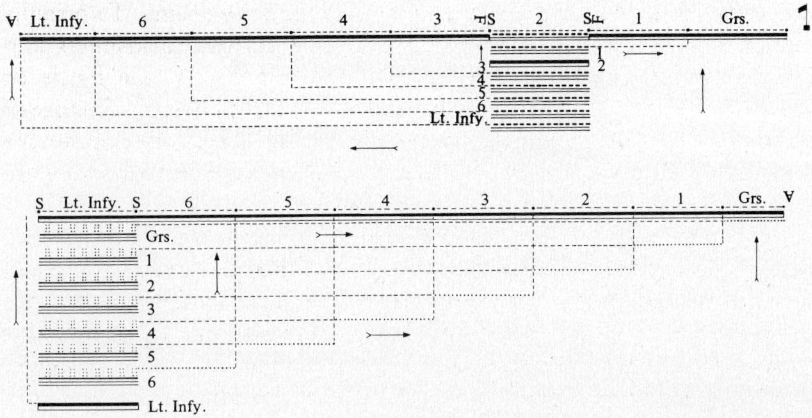


Fig. 4 a and b *A Close Column Deploying into Line on a Central Company* and *A Column at Quarter Distance (Right in Front) Deploying on the Rear Division.* From *Field Exercise and Evolutions of the Army* (1833)

Their ideas formed a bridge to continental practice. The maniple, by combining mass with flexibility and firepower, enjoyed the attributes of the small, shallow columns favoured by the best of the French theorists. For in Europe armies persisted in using the column as the principal means of attack. The column brought an immense weight to bear on one part of the enemy line, but its application depended on several variables. In broken ground, it was a formation less easily disrupted than the advancing line. Whereas the line shrank as it suffered casualties, the column filled from the rear and continued to present an imposing front. On the other hand, its density and length made it a prime target for artillery. Once broken or disrupted, it was, as the Russians demonstrated at Inkerman, hard to regain control, particularly of the men huddled in the centre, unable to see either the enemy or their officers. In the Peninsula Wellington's adroit use of ground had on occasion allowed him to obscure his troops from the French view, and thus to fire on them while still in the deeper column of manoeuvre rather than the shallow column of attack. Although it was the former rather than the latter that he had defeated, British orthodoxy on the column was derived from this experience. The column was thus caricatured as a formation for ill-trained conscripts, fired with revolutionary zeal. Wellington had shown not only its limited firepower, but also that disciplined professionals were able to check it.

The column, however, gave a more rapid advance. The 1824 *Field Exercise* prescribed that columns were for convenience of movement. The quarter-distance column – that is to say companies lined up one behind the other, with a gap equal to a quarter of a company's front between each – was favoured because in brigade every man was within earshot of the brigadier and it enabled quicker deployment into line or square. At Chobham Seaton preferred open column (where the gap equalled the full front of a company), so that not only was the formation always ready to show its largest front to the left or right, but also it was less exposed to the new rifle than a more compact order.[36] However, if the troops in open column were required to deploy to the front, it took considerably longer, and an attack under fire (admittedly a remote possibility), even in echelon, would have been a very ragged affair.

The kernel of the debate between line and column, which had been continuous since the beginning of the previous century, was the need to find an effective blend of fire and movement. The line represented the former, the column the latter, and no absolute solution was possible. The use of the column for rapid movement in the vicinity of the enemy depended for its efficacy on the timing of the deployment into line. If left too late, each file had to pass with its flank exposed to enemy fire until it reached its appointed position in line (although Charles Napier used a system to enable each file to fire as its front became clear).[37] With a large force, deployment into line became impracticable, as gaps would appear while the troops were realigning.

The logical conclusion was an attempt to combine both column and line – the *ordre mixte* enamoured of Napoleon, where columns of attack alternated with battalions in line. Fitzclarence advocated having the front line in line and the second in quarter-distance column,[38] and this was the formation of the infantry at Chillianwallah and Gujerat. Jomini thought an attack best delivered by two lines of battalions, each battalion being in columns of divisions. (A division – in this context – consisted of two companies. Each division would form a continuous line, giving, in an eight-company battalion, a column four deep.) Such columns would be less exposed to artillery fire and were sufficiently shallow to remove the problems of disorganisation and reforming inherent in larger columns. Captain F. Robertson, writing in the *Aide-Mémoire*, proposed that the first line of battalion columns should advance and deploy at the double, covered by the light companies; as they started firing, the second line of columns should pass the first line through prearranged intervals to the attack.[39]

This last manoeuvre, apart from smacking of the flexibility, high training and formalised divisions of the Mitchell–Murray proposals, suggested one way of simultaneously maintaining fire and forward impulse. In the days of smoothbore muskets, the exposed area to cross

Fig. 5 *Battle of Chillianwallah: the Charge of Her Majesty's 24th Regiment through Jungle and Water on the 13th January 1849*: this shows not only the 24th advancing without firing, but also the difficulty of retaining formation over broken ground. Engraving from a drawing by Lt. J. H. Archer, attached to the 24th Regiment (National Army Museum)

before coming to grips was about two hundred yards, enough for the enemy to fire two volleys. One expedient adopted against the Sikhs was to fall to the ground fifty yards from them, in order to let the final and most destructive rounds pass overhead. The conventional view, however, was that nothing should interrupt the flow and unity of the final assault. At Ferozeshah, Hardinge ordered the infantry 'not to fire but to take the position with the bayonet & go up at a walk'.[40] Gough agreed with Hardinge that 'firing before storming caused a greater loss of men', as those who paused to reply were longer exposed to enemy fire. Charles Napier cited with approval Frederick the Great, who required his men to shoulder arms before advancing, to prevent their firing.[41] At Chillianwallah, it was alleged that the 24th had received specific orders not even to load.

The implication in the desire to close as rapidly as possible was that the principal weapon of attack, if not of defence, was the bayonet. 'No troops', Charles Napier declared with the characteristic family aplomb, 'stand a charge of bayonets, and whoever charges first has the victory ... Firing is a weapon ... of defence, not of attack'.[42] The Napierian view saw morale as the essence: 'It is the stout heart and not the *weapon* that gives victory.'[43] Logic told his generation that the French arms in the Peninsula had not been inferior, and thus discipline, physique and courage were deduced as the peculiarly British factors for success. For these the bayonet was the means of expression. Gough's famous cry of relief when told that the artillery had run out of ammunition at Sobraon – 'Thank God! then I'll be at them with the bayonet!' – was an indication of his own faith in the bayonet as 'that never-failing weapon' and 'the proper weapon for the British soldier'.[44]

However, John Mitchell reckoned the belief in the decisiveness of the bayonet to be an idealised notion based on the flowery ellipses of generals' despatches. 'What will posterity think of our bayonets?' he asked.

> Will they ever believe that such rickety zig-zags were ever meant to be used in mortal combat? and what idea must future generations form of the historians and dispatch-writers who have gravely ascribed victories to the power of such weapons? What, again, must be deemed of the military intelligence of an age which could tolerate the tactical puerilities founded on the presumed use of a toy which had been brandished with bombastic fierceness for upwards of a century, and has never yet in fair and manly fight inflicted a single wound on mortal man?'[45]

The riposte to Mitchell was immediate and strong, particularly from William Napier, but when the rhetoric cleared away only three specific fights with the bayonet were established as having occurred in the recent wars.

To fill the vacuum he had thus created, Mitchell was not advocating the end of hand-to-hand fighting but instead – once again aping de Saxe – a reversion to the pike and thrusting sword. He envisaged his maniples divided equally between pikemen and musketeers, the one defending the other according to the armament of their opponents.[46] A pike was a much handier fencing weapon than a musket and bayonet, as the British infantry found out in 1841 when faced by long Chinese spears at Canton. The case for the sword was less clearly established. At the storming of Ghuzni in 1839, the sword was found to have inflicted more wounds but mostly on the limbs, whereas the bayonet had been aimed at the trunk and thus proved more fatal. Furthermore, the high proportion of officers wounded suggested that, as they carried swords, the bayonet was better for self-defence.[47] The opponents of the sword, who included Charles Napier himself, felt that Mitchell, along with the Baluchis, had finally been defeated on this point at Miani. In their view, it was 'perhaps the last great triumph of the British bayonet'.[48]

However, Napier never understood the true nature of his victory. Both at Miani and in Wellington's Peninsular battles, disciplined firepower had been the agent of disorganisation, the onset with the bayonet its confirmation. As Captain Kincaid, the most balanced of the *United Service Journal* correspondents, had pointed out, the bayonet could not be viewed in isolation; it was an adjunct of the musket and its demise could only herald the introduction of a whole new system of weaponry.[49] Mitchell's primary contribution was a more realistic view of the bayonet's capabilities. It was acknowledged that bayonet fights had rarely occurred, and that its effect was a moral one, its onset precipitating the break-up already developing in the enemy ranks. Mitchell's revisionism prompted an analysis of the occasions when the bayonet did serve a useful purpose, that is to say when hand-to-hand fighting occurred – in surprise and night attacks, and also in confined country.

Acknowledgement was nonetheless reluctant: the bayonet was a badly made weapon inexpertly used. It was necessary because infantry had to have muskets. Kincaid therefore brought the final logic to an argument dominated by extremes: the true answer to Mitchell's problem lay in the proper training of the infantry in the bayonet's use.[50] So unsure was the soldier of the weapon in which he was alleged to have such confidence that at Merthyr the 93rd Highlanders with bayonets fixed were defeated by Welsh dissidents armed with staves.[51] A few attempts were made to formulate a bayonet exercise: Sir Henry Torrens drew up one of the first but it led to ruptures and 'attitudinising'.[52] Henry Angelo, the superintendent of sword exercise, prescribed a system which enjoyed some currency, but he seems to have relied on occasional personal instruction to propagate his method. However, in 1849 he produced in book form *Angelo's Bayonet Exercise* and, although Richard Burton, the Arabist, entered the field in 1853, Angelo's system was adopted in several regiments. At the Cape the 74th found its regular training in the use of the bayonet most valuable.[53] Regardless of whether the bayonet was genuinely effective or not, the principal consequence of practice in its use was confidence in its efficacy on the part of those who were equipped with it – so much so that the men preferred to employ it rather than rely on an ill-developed firepower. However, the Horse Guards remained suspicious, having queried the reason for a bayonet exercise at all in 1850.[54] Angelo's book was submitted to Hardinge in 1853,[55] but did not secure official approval until 1857.

The slow development of the bayonet exercise was partly attributable to the construction of the weapon itself. In addition to the ease with which the blade bent, the system of attaching it to the musket was insecure, soldiers having their bayonets pulled off the muzzles during riots in Cork[56] and, more seriously, in the fight at Miani.[57] Realistically the 1824 *Field*

Fig. 6 *The 31st Regiment Advancing to the Charge at the Battle of Moodkee, on the 18th of December 1845*: the bayonet applied in the manner approved by Gough. Engraving by J. Harris after H. Martens (National Army Museum)

*Exercise* did not require bayonets to be fixed during firings and movements, but the 1833 edition revised this ruling with the result that after the manual exercise 'there was always a wild scramble to recover those which had flown off or stuck in the ground during the parade'.[58] A spring was required to hold the bayonet in place, but that attached to the Baker rifle had to be unbolted before firing could commence, with the result that in nine out of ten cases the bayonet fell off.[59]

In 1844, George Lovell, the superintendent of small arms, although opposed to shooting with bayonets fixed as it upset the balance of the musket, developed a new weapon. The bayonet was fixed with the customary zig-zag motion but was held in place by a spring-catch under the muzzle that hooked over a ring on the socket: in trials Lovell's bayonet was still firm after fifty rounds.[60] Improvement it undoubtedly was, but perfection it was not. In practice the spring was either too stiff or worked loose, for in one company's ball firing, fifty bayonet springs were broken and a further twenty-five bayonets came loose.[61] However, the adoption of the Enfield rifle in 1853 wrought a great improvement in the manufacture of the weapon. The barrel of the rifle was shorter than that of the musket, and yet to retain the overall length considered desirable for infantry to defend themselves against cavalry would have meant such a weight of iron in the bayonet that the imbalance would have rendered accurate shooting difficult. Consequently the bayonet was made of steel, was five ounces lighter than its predecessors, and was held in place by a locking ring.

Despite the improvements in its use and construction, Mitchell's relegation of the bayonet to a minority position was not challenged. At Inkerman, when allegedly bayonet had indeed clashed with bayonet, only six per cent of wounds were caused by the 'rickety zig-zag'.[62] Inkerman was of course a defensive fight, and confirmed that the strength of the infantry lay principally in firepower. As Wellington showed repeatedly, a penchant for the attack was no argument against disciplined musketry.

Perhaps a principal reason for the undue admiration of the bayonet was the imperfection of the musket with which Wellington's army had been equipped. The Peninsular arm was the celebrated Brown Bess, a smoothbore flintlock musket, with a calibre of .753 in. It weighed almost twelve pounds, and even after a short period of drill Ensign Chichester found it 'terribly heavy & weighed on my knees & ankles & feet very much'.[63] A lighter musket, however, would have increased the fearsome recoil. It was argued that the calibre could not be reduced, since the bigger the ball the better it overcame resistance and crosswinds. Moreover, a large bore (the largest in service in Europe) allowed all calibres of ammunition to be used and meant that the size of the round inflicted more ferocious wounds.

Wellington himself considered

> our Arm as the most efficient that had yet been produced. The fire from it undoubtedly and acknowledged to be, the most destructive known. It is durable, it bears all sorts of Ill-Usage; is easily repaired, and kept in Repair and Serviceable; and besides its Power as a Missile, its length is an advantage in the use of the Bayonet.[64]

Wellington's practice of positioning his line below the crest of a hill gave weight to his faith, since the fire was then held until the last moment and the full punishment extracted by the massive ball. The musket therefore was simply 'for the use of numbers against numbers, in close order', and its excessive windage (0.07 inches as against 0.039 inches in France),[65] although contributing largely to its inaccuracy, was off-set by the high rate of fire (three rounds a minute) that the ease of ramming down thus enabled.

Trials showed that a great improvement in accuracy was possible with a well-wadded round to reduce the windage, but the vagueness regarding the Brown Bess's actual capabilities reflects the rarity with which it was ever put to the test at even moderate distances. Harry Smith reckoned that it ranged eight hundred yards,[66] and William Napier said he had seen a man shoot an individual French officer at three-quarters of a mile.[67] However, experiments conducted in 1825–6 showed that from 66 to 70 per cent hits at a hundred yards, the capability of the Brown Bess dwindled to 25 to 30 per cent at two hundred yards, despite an increase in the height of the target from three to five feet.[68] A modern analysis is even less flattering: in the Seven Years War and in the Peninsula 2 per cent of shots were effective, although this might have risen to 5 per cent at ranges of a hundred yards or less.[69] Therefore most manuals by the 1840s reckoned two hundred yards was the maximum effective range of the musket.

It was only just as the Brown Bess was nearing the end of its long life in the British army that any definite idea of the weapon's performance was secured. In 1846, on Sir John Burgoyne's instructions, the Royal Engineers conducted some experiments 'with a view to improving the very limited knowledge at present existing, as to the range and power of the musket'. At point blank, the range of a single musket over thirteen shots varied from 116 yards to 218, and, at ½° of elevation, in ten shots the variation was 126 yards to 300. At 50° elevation, an average range of 645 yards was secured, but this could hardly be construed as accurate fire when it involved aiming 130 feet above the target. The lateral deviation was reported as even greater than the vertical. From such data, conclusions were hard to draw, but 'as a General Rule musketry should not be opened at a distance exceeding 150, and certainly not exceeding 200 yards, as at and beyond that distance it would be a mere waste of ammunition to do so'. At two hundred yards the pattern of fire set up at one hundred was not maintained, half the shots missing the target, some falling thirty yards short, and others passing to right and left.[70]

The report did not so much clarify the position as condemn the musket. The Horse Guards' staff poured cold water on it.[71] However, the Master General of the Ordnance and Lord Frederick Fitzclarence were impressed by its results,[72] and Major J. Bentham took care to publish its findings in the *Naval and Military Gazette*,[73] as well as sending copies to Lord Grey and to influential soldiers such as Lord Cathcart, William Eyre and Lord Seaton. It was therefore to lay the ground for the adoption of a new and vastly superior weapon.

The principal reason adduced for the enormous fluctuation in performance was the variation in size, density and manufacture of the ball. Furthermore, no variation was made in the size of stock to fit the firer, the triggers were too stiff, and no backsight was fitted. 'In nine cases out of ten', Mitchell wrote,

> the difficulty of pulling the trigger makes the soldier open the whole of the right hand in order to aid the action of the forefinger; this gives full scope to the recoil:

the prospect of the blow makes him throw back his head and body at the very moment of giving fire; and as no aim is ever required he shuts his eyes, from the flash of the pan, at the same instant, so that the very direction of the shot becomes a matter of mere accident.[74]

The first attempt at a major revision of the British infantry's armament since the days of Marlborough was initiated in 1835 with the appointment of Sir Hussey Vivian as Master General of the Ordnance. Vivian felt that with practice all soldiers could be good shots and thus their morale and confidence on the battlefield enhanced. The problem was the musket itself, and in 1840, to remove its principal manufacturing defects, he selected George Lovell as Inspector of Small Arms. Lovell's task was to supervise the production, proof and supply of all firearms, bayonets, and swords.[75] In 1837 Lovell had recognised the inability of the Birmingham small arms industry to produce standardised weapons, with identical and interchangeable parts. The French ideal of government factories contrasted favourably with the cottage industry in Britain, which had little capacity for expansion in war. Realist as he was, Lovell saw no prospect of a wholly state-run arms industry and confined his suggestions to the resuscitation of the Ordnance establishment at Enfield so that it should 'maintain a full and certain check upon the prices and workmanship of the Trade'.[76] This, with Vivian's support, he achieved. Technological limitations prevented the degree of standardisation Lovell aspired to, but nonetheless Greener, the famous London gunsmith, complained of the high levels set by the government. 'Those small specks in a barrel are noticed; a lock plate or main spring being the shade of thickness of a piece of tissue paper out of size; probably it may be a little too broad, and not of the required height. You could scarcely credit what trifles a musket, and the various parts of it, are rejected for.'[77]

Once the musket had passed to a regiment, Lovell secured its better care by ensuring that each armourer sergeant underwent a probationary course of instruction at Enfield and passed an examination in his trade. Lovell himself penned an official manual, *Suggestions for the Cleaning and Management of Percussion Arms* (1842), that acted as the armourer's vade-mecum. After 1847 every musket had to be inspected once a month, although this was too great a task to be rigidly applied.

The particular reason for Lovell's concern was his introduction into the British service of a more sophisticated military firearm, and in particular of his own pattern of percussion lock. The flintlock had a sluggish discharge, often hanging fire for several seconds, and its flash tended momentarily to blind the firer, on occasion burning his face or hair. In one set of trials, 1,448 rounds out of 27,000 had misfired.[78] The problem was exacerbated by the rain: after a slight shower, only four muskets in a whole battalion went off,[79] and in China in 1841 this failure occurred in action. These were faults inherent in the flintlock, but in addition the Brown Bess's spring was too strong for its flint and hammer, and the arm was thus rapidly rendered unserviceable.

Trials and inventions relating to the percussion system had been current since the work of a Scots minister, the Rev. Alexander Forsyth, at the start of the century. Lovell himself entered the field in 1824, and won the support of the Master General of the Ordnance in 1830.[80] A series of methodical trials proceeded, with Lovell endeavouring to answer three sets of questions – mechanical, chemical and economic. Inevitably it was the latter consideration that dictated the first pattern of percussion musket, which had to be a conversion to use up the existing store of flintlocks. In 1836, four main competitors were selected and sent for regimental trial: the results favoured the converted flintlock, although, with the exception of its enclosed back-action lock, Lovell's own arm was highly praised. The former registered 1,280 hits and six misfires out of 1,520 rounds; the Brown Bess, also included in the trial, scored 992 hits and, more seriously, hung fire forty-five times and misfired on 167 occasions.[81] It was therefore certain that the percussion system would be adopted, and a visit to France by Lovell in 1838 confirmed his own predilection for a copper cap as its primer.[82] Further trials seem to have been envisaged, as reports were awaited on the performance of the percussion musket from Canada. But the revelation that the store of arms had been reduced from 722,000 in 1829 to 109,000 meant that in 1839 the converted musket was adopted as a stop-gap.[83] In consequence Lovell's back-action lock was only adopted by the Guards, although a lighter musket with a side-action lock was approved and became known as the 1842 pattern.

The decline in arms stores revealed in 1839 was compounded in 1841 by a fire at the Tower. Many of the existing stocks were destroyed, including newly set-up percussion arms. The issue of the new muskets was therefore very slow, and their benefit consequently muted. In response to Lovell's concern about the whole problem of weapons procurement, Vivian had agreed in 1837 that a reserve of 50,000 of the eventually approved muskets should be established before any regimental issues were made.[84] Unless arms were distributed in sufficiently large batches, troops might find themselves at war with different weapons, and therefore require varying types of ammunition. The likelihood of this occurring was diminished in 1837, when Lovell secured a uniform calibre for all weapons in the British service.[85]

Nonetheless things were little better by 1843. In that year, Hardinge roused Peel, the Prime Minister, to the dangers facing Britain should invasion occur and the militia have to be called out: for such a contingency a store of 250,000 stand was required, but only 60,000 were available.[86] Vivian's successor, Sir George Murray, was continually badgered by Peel on the need to build up good stocks, and, by restricting the issue of new percussion arms, Murray had by 1845 created a pool at home of 34,952 percussion muskets and 57,444 flintlocks, with a further 3,350 and 106,549 abroad.[87] However, the Marquess of Anglesey seems to have been the first Master General to see the weakness of the whole system: frantic about home defence, he confided to the Duke in 1846 that 'The present means [of supplying firearms] can only produce what is required, at a fearfully distant period.'[88] With less than half the projected store of 200,000 percussion muskets achieved by 1849,[89] Anglesey was in 1851 driven by the failure of the Birmingham gun trade to conclude that 'a great [government] establishment for the manufacture of our own arms' was the only solution.[90] All that Lovell's revival of Enfield achieved was to establish higher standards of workmanship: it did not yet manufacture to any great degree itself.

This failure in arms procurement was a severe check on the army's efficiency. The first percussion arms were issued to troops in the United Kingdom, largely because they were most likely to be involved against a European enemy, whether fighting invaders or as an expeditionary force abroad. Thereafter, however, Murray's deliberate restraint meant that the adoption of percussion muskets was slow. In 1847 all infantry who had not yet received percussion muskets were to do so,[91] but even in 1849 the 96th Foot still had flintlocks in anticipation of their going to India, where the East India Company would have to pay for their replacement.[92]

The initial response from most regiments to the new arm – when they eventually got it – was extremely favourable. Vivian's primary reason in sanctioning its introduction was to ensure good performance in wet weather, and in this he was utterly vindicated. Lovell's concern for the quality of workmanship resulted in much greater attention to shooting among the men, who began to take pleasure in what had been a painful chore.[93] The introduction in 1839 of Napier's compressed ball machine and the adoption of a backsight were perhaps the principal causes of more consistent practice at ranges of up to 250 yards. The main doubt had been that the rate of fire would be slower than with the flintlock, particularly when placing the cap in action or in cold weather. In the event, although at first the practice was slower, the old rate of three rounds a minute was regained. The fact that there was no flash from the pan of the neighbouring flintlock when firing in line meant that accuracy rose by a third.[94] Reports as late as 1852 proved that 'these arms have not shown a tendency to give way in any particular', the defects mostly being minor and easily repaired.[95]

Tributes for the percussion musket, however, were no more than an indication of what a properly made Brown Bess might have been. As one authority has written, 'The basic part of Lovell's musket, the relationship between barrel and ball, was, after all, no different, except in finish, from the first model Brown Bess.'[96] The criticisms about the windage, the trigger pull, the excessive recoil and the limited range were still all applicable. The difficulty of drawing the priming if it had become damp overnight rendered it objectionable in the field, and by contrast firing twenty or thirty rounds in battle made the musket too hot to hold. Greener, who admittedly was embittered by the rejection of his own weapon, maintained that the friction necessary to bring the main spring to full cock was so great that in many cases the blow onto the cap was lessened and had even led to three or four misfires before the cap would explode.[97] Above all, the musket remained as inaccurate as ever. In other words a rifled barrel, or at the very least considerably reduced windage, was required.

The principal reason for not adopting the rifle was its slow and difficult loading. The rifling and the tight fit of the ball meant that the Baker rifle (the arm carried by the rifle regiments from their inception until 1839) could only maintain a rate of fire of a round a minute, and very often its user had recourse to the mallet with which he was provided to drive the ramrod down. Nor was the accuracy imparted by the rifle sufficient to justify its wholesale adoption in the army. The Baker rifle averaged only a third hits at 300 yards on a six-foot high target.[98] Taken in conjunction with the belief that perhaps one man in ten could ever be a competent shot, the risk in the universal issue of the more sophisticated arm was not felt to be justified.

Lovell himself subscribed to this view. The rifle required such careful loading that

> I am entirely satisfied that a Rifle is not adapted for Battalion Service; and is not so effective for such use as a plain Barrel, correctly bored. The Rifle if continued at all in the Army should be put into the hands of only a few picked shots in each Regiment, and the Arm ought to be constructed and loaded upon the most perfect Principle for accurate shooting at long Ranges; to be resorted to only occasionally in Sieges and against reconnoitring Parties etc.[99]

This need for snipers with more sophisticated weapons was a recurrent one that still has a modern application. It was ill-served by there being only four battalions in the British army equipped with rifles; obviously they could not be in every part of the globe at once. Furthermore these battalions recruited in the same manner as the rest of the infantry, and were not composed of picked marksmen. Assuming that the rifle was a specialist arm, the Austrian allocation of a battalion of riflemen per brigade and the arming of the third rank with rifles was a fairer distribution. In Canada, in 1840, Sir Richard Jackson suggested arming six men per company in all regiments with the rifle, and surprisingly the Horse Guards approved.[100] The consequent success of the 93rd with their new arms pointed to the potential application of rifles to all infantry.[101]

Lovell's belief in the limited application of the rifle to warfare suggests that his contribution to its development would be restricted. In 1836, he had been working on a percussion rifle to replace the flintlock Baker rifle. However, in that year, a rifle was sent from Brunswick with only two grooves, cut deep on either side of the barrel. Lovell was convinced of its merits: the simpler grooving suggested the easier loading associated with smoothbores; an eleven-grooved weapon needed cleaning after two or three rounds, but the Brunswick fired as well in trials after fifty rounds as at the start; at 300 yards the two-grooved rifle put nineteen out of twenty shots into a six-foot square target. The Ordnance Select Committee was equally impressed: in particular the Brunswick could fire ten rounds of the belted ball (the ball had a ridge on its circumference designed to fit into the grooves) in seven and a half minutes and ten conventional rounds (albeit with diminished accuracy) in four and a half minutes.[102] The size and strength of the grooves meant that a larger charge could be used, with the result that greater accuracy was achieved at long ranges. Although sighted to 200 or 300 yards, accounts exist of accurate shooting to 400 or even 500 yards.[103] Amidst the customary plaudits the Brunswick percussion rifle was approved in 1837, and issued to the rifle regiments in 1839.[104]

The early satisfaction turned rapidly to disgust. The weapon was not well made, and deteriorated with use; at the Cape the 1st Battalion of the Rifle Brigade found the woodwork unable to take the shock of concussion, with the result that there were 211 broken stocks in five companies.[105] Much of the criticism centred on the belted ball and its two grooves, and, since in the event the barrel started fouling after only a few rounds, it became jammed in the quick firing required in action. As he became more flustered in the heat of battle, the greater was the likelihood of the soldier's inserting the ball slightly on one side, which meant that he either had to turn his rifle upside down to start again or that the force needed to ram down the ball worked away the edges of the grooves. Not only did the effort render a man's hand too unsteady for accurate shooting, but also the consequently loose fit of the ball meant that it rolled in the direction of the grooves, thus diminishing the effect of the rifling. The distinctive ball was itself often bruised in the ammunition pouch by the motion of the wearer. The Ordnance Select Committee had recommended that a third (i.e. twenty rounds) of the rifleman's ammunition should be plain ball, so that he could if necessary increase his rate of fire, but in practice only ten such rounds seem to have been carried[106] and their use in any case amounted to an admission of the defects of the Brunswick.

Much of the vituperation loaded on to the Brunswick as a rifle, as distinct from its durability as a weapon for the field, stemmed from the higher standards being attained on the continent. The French, in trials of 1844 and 1847, had established that their 1842 pattern rifle was marginally more accurate at 200 metres and twice as accurate at 300.[107] Even more impressive were reports from Belgium and France of the *carabine à tige*, which took its name from the pillar in the breech designed to drive the bullet so that it would expand into the grooves of the rifling. The problem of windage was thus removed, while loading became if anything a simpler operation. 'It is alleged', Sir Frederick Smith reported to Fox Maule, the Secretary at War, in 1850, 'that this rifle throws its bullets with as much accuracy at 1000 French metres, as the old rifle did at 600 metres, and that the bullet of the former at the great distance, has more penetration, than that of the latter at the shorter distance.'[108]

However, rate of fire rather than long range remained the dominant official consideration. Therefore, much more than the *carabine à tige*, it was the news in 1849 of a Prussian needle-fire, breech-loading rifle that aroused interest and speculation. If technologically feasible, the breech-loader promised a much greater rate of fire, the solution to all the problems of windage and ramming down, and a system of loading that was possible on horseback, behind cover or lying prone.

In 1850, the government addressed itself seriously to a reconsideration of its military firearms. By March, reports from abroad had been collected, including one on the Dreyse, the Prussian weapon. In the following month the recently constituted Small Arms Committee was convened with the express purpose of considering the new arms of France and Prussia, and also the Lancaster rifle, recently tried at the Cape by the Rifle Brigade.[109]

The result of the Committee's trials went firmly against the Dreyse. 'Whatever be its merits in other respects, the needle-prime musket is too complicated and delicate an arm for general service.'[110] It had easily fired six rounds a minute, and maintained accuracy to 800, or even 1,200 yards. However, particularly when the gun was heated and foul, the operation of withdrawing and replacing the bolt required much strength. The barrel tended to wear at its junction with the cylinder, and related to these two problems was the perennial difficulty with breech-loaders, the escape of gas at the breech that grew progressively worse as firing continued. The spring that drove the needle was delicate, and in addition when the needle was dirty it tended to misfire. Colonel Hawker considered that a new needle was required every twelve shots.[111] British

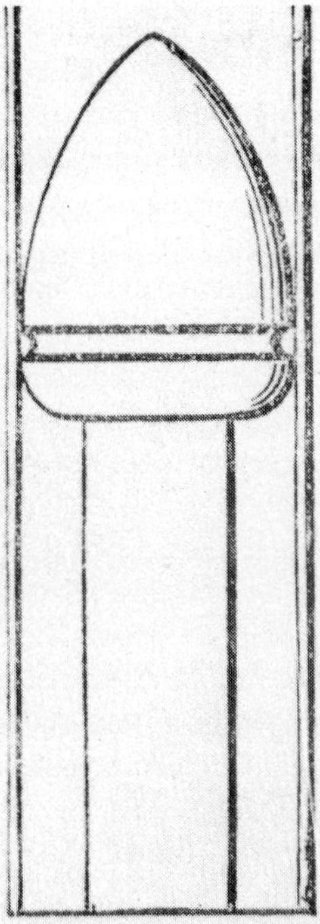

Fig. 7 A cross-sectional view of Thouvenin's pillar, at the bottom of the rifle barrel, with Delvigne's cylindro-conical bullet. From J.-W. Jervis, *The Rifle Musket* (1854) (by permission of the Syndics of Cambridge University Library)

doubts were reflected in Prussia, where they only shot at up to 600 paces, and confessed that the fire of the *carabine à tige* was superior.

Nonetheless, it was widely felt that, if the difficulties could be overcome, the breech-loading principle should be adopted. Sir Howard Douglas, who had condemned the Dreyse, urged the fullest trials on breech-loading rifles.[112] Although fearing their adoption might lead to a waste of ammunition, Hardinge remained interested in such weapons and in 1852 Sir James Simpson confessed to him that he believed all the members of the Small Arms Committee wished to see the breech-loader eventually adopted.[113] However, by 1854, none of the solutions offered had overcome the problem of escaping gas. The mechanical deficiencies meant that the principle of breech-loading was not to be accepted in the British army for another decade.

Therefore, despite the rapid fire of the Dreyse, the 1850–1 trials favoured the French rifle. Although basically this was the pillar-rifle accredited to Thouvenin, it had one modification, worked out by the Frenchmen Delvigne and Minié – an iron cup was inserted in the hollowed base of the conical bullet. Thouvenin's simple pillar had in many cases required to have the bullet rammed onto it, causing subsequent unsteadiness in the firer's aim. Delvigne's and Minié's refinement made the cup expand the bullet into the rifling and rendered the so-called Minié rifle a practicable military arm.[114] The other feature of the bullet was its conical shape, that not only offered less resistance to the air and penetrated further on arrival, but was also reckoned better adapted to the rifling than the conventional spherical ball. As Lieutenant-Colonel A. Gordon remarked, 'the whole discovery' of the Minié rifle was in the bullet.[115]

A tour of Europe by Lovell revealed no fresh rival to the Minié, and in May 1851 Wellington accepted Anglesey's recommendation that all the infantry should be armed with it.[116] Such a dramatic leap in the Duke's thinking has represented too much of a volte-face for subsequent historians, and it seems that Wellington was so shocked by his own temerity as to reverse his decision. However, Anglesey, while recognising that in the current state of technology and European military concern the Minié could only be an interim solution,[117] to his eternal credit insisted that the entire force receive the Minié, as the Small Arms Committee had recommended.[118] The approved weapon had four-groove rifling, which made one turn in 6 ft 6 in. In order to be uniform with existing service calibres, the bore was .702 in. This means that sixty rounds and seventy-five caps weighed over seven pounds, when the total weight of the rifle and bayonet was only just over ten and a half pounds. Reflecting the expectation that an improved rifle would soon offer itself, only 28,000 were ordered, although to ensure uniformity of ammunition in the Crimea further orders meant that 34,000 were made.[119]

The initial 500 were ready for regimental trial in October 1851.[120] On Grey's insistence,[121] the first actual issue was made to troops at the Cape, so that their capabilities on service and in a hot climate could be properly assessed. Accordingly six marksmen per company received them in April 1852.[122] But because of the failures of the Birmingham contract system the complete order was not fulfilled until November 1853. Therefore, so as to give as much opportunity as possible for all ranks to familiarise themselves with the new arm, in July 1852 a scale of only 100 stand per regiment was established. In January 1853, each company of infantry at home was to have twenty-five Miniés, and the rifle regiments were to adopt them throughout.[123] The logic of not as yet issuing them generally abroad was justified when the home army had to mount an expeditionary force the following year: all regiments going to the Crimea were to have the 1851 pattern rifle-musket,[124] and it was the Minié that was the main infantry arm on the uplands of Sebastopol.

A Russian staff officer spoke of whole regiments 'melting' before the British fire at Inkerman.[125] In particular it was the penetration of the Minié that earned for it such a formidable reputation: 'These Minie bullets', one officer reported from Balaclava, 'do not make a practice of stopping half way in a fellow; some in fact have been seen at 400 yards to go clean thro 3 or 4 Russians and kill them all.'[126] In trials the rifle's round had penetrated 5.25 inches into a deal target at 250 yards; that of the musket penetrated 2.875 inches.[127] An unfortunate drummer marking shots behind an earth butt four feet wide, tapering to two and a half feet at the top, was shot dead by men firing 300 yards away.[128]

Coupled with this was the Minié's ease of use and reliability. Of 17,739 issued up to March 1854, only nine suffered burst barrels.[129] In a series of trials conducted by the expeditionary army at Varna in August 1854, a rate of fire of over two rounds a minute was easily maintained. Not one stoppage occurred and, after the first twenty rounds, loading in fact became easier as firing cleared the fouling.[130]

However, the greatest single attribute of the Minié was its range. It was said that Minié himself had achieved reasonable practice at 1,804 yards, and the Russians claimed that it took effect at 1,600 yards. Eight hundred yards was the accepted distance up to which the arm's fire remained certain. At this range, a score of 77 hits out of 100 had been registered on a target eight foot square.[131] At 150 yards the Minié was twice as accurate as the smoothbore musket, and at 400 eleven times better; at ranges between 300 and 600 yards it was calculated that the execution done by a hundred and fifty men armed with the Minié was equal to that of five hundred and twenty-five equipped with the musket.[132] Furthermore this superiority of practice was not confined to a few marksmen: at 300 yards, thirty-three of fifty-nine unselected men put seven out of ten shots into the target, and only twelve scored less than five.[133]

It was clear to the Minié advocates, as much as to the opposition, that perfection would not be attained in one bound. However, the defects were comparatively minor. Most of them centred on the ammunition. The iron cup tended to blow through the bullet, on occasion leaving lead in the barrel. It was soon realised that the bullet would travel as well without the cup, but even then it was said that the head was forced off, leaving the sides of the hollowed cylinder in the ball. Since it was a conoidal ball, it was reckoned that it had no side to rest on, and thus tended to tilt forward in a way that a cylindro-conoidal bullet would not. More important was the question of ammunition supply. Wellington had insisted that the Brown Bess's calibre should be preserved.[134] Thus only fifty rounds instead of the usual sixty could be carried, but at the same time the improvement in the weapon encouraged greater expenditure in ammunition, particularly since the enemy could be engaged at longer ranges. As we have seen, the logic behind his thinking was that a larger ball travelled further and faster, and hit harder. In addition all the army's weapons could use the same ammunition, as well as those of the smaller calibres employed by the other European armies. However, the superior penetration of the Minié suggested that while the weight of the Brown Bess ball might be retained, its calibre in a conical bullet was a superfluous burden.

Another set of problems related to the sighting of the rifle. Not only was the graduation inaccurate, but also the slide tended to become loose and consequently slip on each discharge.[135] The shortest range for which it was sighted was 200 yards, and at 100 yards the Minié therefore aimed between two and three feet high. This was easily overcome by the soldier aiming low, and in fact so good did his judging of distances become that in the flurry of action he tended to dispense with the sight altogether.[136] Nonetheless doubts remained about the Minié's performance at short range.

These criticisms were to be eradicated easily by a series of trials, but in the short term they provided plausibility for otherwise specious arguments against the adoption of the rifle. Sir Howard Douglas reckoned that the Minié's long range could only be achieved by a great trajectory that demanded pin-point accuracy. He calculated that at 1,000 yards, the highest range for which the Minié was sighted, the elevation required would allow a longitudinal deviation of eight yards, 'a degree of precision which it may be presumed will never be attained'. The common musket aimed at 300 yards would command a greater proportion of the intervening ground. In fact Douglas's scientific reputation was being used to mask his bigotry: he believed that 'Rifles will never answer for line-firing', since their mechanisms, ammunition and sights were too complicated for the average infantryman, who was inherently incapable of firing properly.[137]

Douglas's justifiably high reputation in the field of artillery leant weight to his case. Fox Maule, the Secretary at War, opposed the universal issue of the 1851 pattern on the grounds not only of expense but also of efficiency.[138] The Rev. G. R. Gleig, the Chaplain General and in other respects an ardent army reformer, was so convinced by Sir Howard's arguments that he branded the Minié as 'certainly one of the clumsiest and most cumbersome weapons wherewith an infantry soldier was ever loaded'.[139] Even the progressive editor of the *United Service Magazine* was moved to claim that in this respect 'we are no advocates of innovation' but belong to the 'old-fashioned, practical reflecting few'.[140]

The qualms of all these critics revolved round the traditional and restricted use of light infantry and the dominance of the line and column. But the tactical orthodoxy on which such thinking was based, the rigid division of line and light infantry, the solidity of one and flexibility of the other, was already, particularly in the British context, wavering. Even Douglas himself was caught midway between conservatism and admiration for the rifled breech-loader. Gleig confessed that his arguments were quashed by the improvements to be rendered in the 1853 pattern rifle.[141]

The most determined and blinkered attempt to stop the flow came from, in a sense, a surprising quarter. Sir Charles Napier was a long-standing protagonist of the musket, which with more careful construction and a lesser weight he conceived to be the simple weapon desirable for use in battle. On the shaky foundations of his ignorance of the Minié's capabilities, by which he endowed it with the failings of earlier rifles, Napier constructed an edifice that reflected his own primary concerns, which were for the soldier's morale and happiness. 'Minié rifles . . . will in my opinion destroy that intrepid spirit which makes the British soldier always dash at his enemy', since 'The short range and the very uncertain flight of shot from the musket begets the necessity of closing with an enemy.' Napier envisaged the adoption of the rifle leading to long, drawn-out battles, fought with skirmishers and consequently higher casualties. The traditional British strength of moral and physical courage would be forfeit. 'These new weapons turn our soldiers from confidence in *themselves* and their commanders, to confidence in their weapons.'[142]

With regard to tactics, Napier's prognosis was not totally inaccurate. On the subject of morale he was proved too pessimistic: Estcourt, the Adjutant General in the Crimea, reported from the Crimea that the men's interest and faith in their new arms had proved a great fillip to their confidence.[143] Furthermore an attempt to halt the progress of technology in Britain alone would be no solution when faced by a better-armed European enemy. Douglas, Napier and the others saw that the new firepower would break the line or the column but were reluctant to conclude that a new and looser tactical formation would have to replace them. It is therefore greatly to the credit of Hardinge, Anglesey's successor as Master General, that he met the opponents of the rifle by removing their specific criticisms, by perfecting the soldier's arm, but not by returning to the smoothbore musket.

Both Masters General were under extreme pressure from the gun trade. Stimulated by the Great Exhibition, officers at the Cape were arming themselves with rifles made by some of the finest sporting gunsmiths in the land – Egg, Purdey, Manton, Deane and Adams, Beattie, Wilkinson, Lancaster and Dickson.

Scarcely a mail-steamer arrived from England without bringing out some new improvement in fire-arms; smooth bore, two groove, four groove, and polygroove, half and quarter twist, rifle barrels; conical balls, plain and winged, sharp-pointed and rounded, with wooden plugs or iron cups, concave and convex bases, etc., were each in turn tested, advocated, or rejected, affording a never-failing topic of interest and conversation.[144]

A spirit of invention and evolution was in the air. 'It is impossible to stand still', Burgoyne minuted, 'and therefore the least imperfect [weapon] must be used until further assured improvement shall be devised.'[145] The Small Arms Committee, reflecting the doubts about the larger calibre of the Minié, in January 1852 recommended a lighter musket should be made, whose reduced ball would take an elongated projectile of the weight of the spherical ball. When Hardinge succeeded Anglesey the following month, both Wilkinson and Lovell were already submitting weapons on these lines.[146]

The new Master General was as feaful of invasion as his predecessor. The low state of the stores of arms and the threat of possibly superior French firepower meant that 'At no period of our history, have we ever been in a state so favourable for change in the pattern of our *Small arms*.'[147] Hardinge's concern for the soldier's armament was such that, when Commander-in-Chief and even during the hectic year of 1855, he as far as possible corresponded personally on the subject: 'There is no part of a soldier's equipment in which the C-in-C takes so much interest as that of possessing, and being able to fire the new rifle ... You cannot tell him too much or be too minute. His great ambition is to bring the arm to perfection.'[148]

Hardinge's immediate step was to instruct 'our most eminent Gunsmiths' in April 1852 to prepare 'Muskets of the Regl length, to carry a ball 1 oz ¼ – rather exceeding the old Musket ball – the gun with its bayonet to weigh 9 lbs 4 but not more – instead of 11 lbs 3 the last pattern for the Line'.[149] It was thought probable that the calibre would be .65 in and that the cup could be omitted from the bullet.

Purdey, Westley Richards, Lancaster, Wilkinson, Greener and Lovell all competed. But the best shooting in the whole trial was by a short rifle made at Enfield. In consequence, a fresh Enfield rifle was made in August 1852 that embodied the experience of the Minié's defects with the best of the gunsmiths' submissions. It weighed 9 lbs 3 oz, and its bore was .577 inches. The rifling was three-, not four-grooved, since it was found that the opposition of grooves and lands caused the ball to be less disturbed and to expand more readily. The bullet, made entirely of lead, was cylindro-conoidal, without a cup but retaining the hollowed base. The sight was that from the Westley Richards rifle, which was graduated to 800 yards and which alone of those submitted could take a blow in any direction without breaking.[150] The result was a weapon that could 'kill at a distance

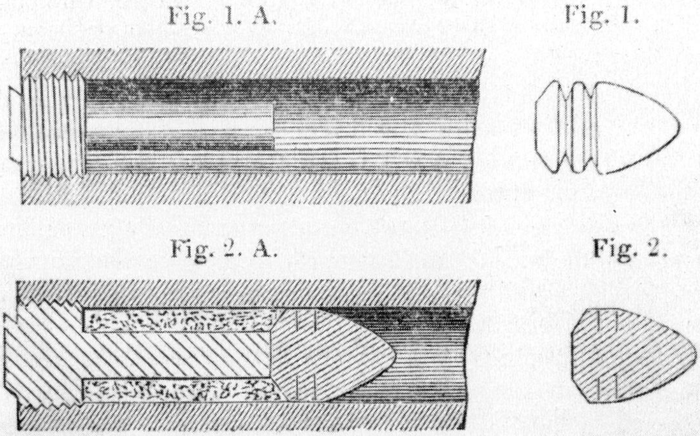


Fig. 8 The pillar and bullet of the Lancaster rifle, before loading (Fig. 1) and after the bullet has been compressed by the force of the ramrod (Fig. 2). From F. R. Chesney, *Observations on the Past and Present State of Fire-arms* (1852) (by permission of the Syndics of Cambridge University Library)

as far as the men firing it can see the enemy', a barrel 'proved to be nearly twice as strong as the barrel of 1842', and ammunition that enabled the soldier once again to carry sixty rounds.[151]

However, in the Enfield's subsequent trials against the rifle submitted by Lancaster, the latter tended to perform better. Lancaster's weapon eschewed conventional rifling, the barrel being cut in an ellipse, the major axis exceeding the minor by .005 inches and thus imparting spin to the bullet. The Lancaster did not strip and had less recoil. The fact that it was in essence a smoothbore meant that the windage was considerably less, resulting in greater velocity and a curve in the trajectory a quarter that of the Minié. Hardinge was duly impressed, and in August 1853 20,000 rifles of .577 inch calibre were ordered to be made, but with the rifling undecided pending further trials.[152] At first the elliptical barrel seemed destined for adoption, but during prolonged firing in December 1853 the Lancaster began to strip and lose accuracy. The three-grooved rifling was consequently adopted.[153]

To describe the pattern 1853, the so-called Enfield rifle, as a finalised weapon would be to ignore the totally transformed spirit of army administration. As an order of merely 20,000 testified, further improvements and modifications were expected. The cocks were found to be too slight and so highly tempered that they tended to break off. The small passage between the charge and the nipple led to a higher proportion of misfires than in the Minié. In particular the ramrod gave trouble; designed with a swell so that the soldier's fingers, made greasy by handling the bullet, could grip it, it tended to expand and stick in hot weather or shrink and rattle in dry.[154] Even when these faults were corrected, Hardinge still awaited further reports from the Crimea: 'I do not think it would be prudent finally to submit it as the Regulation weapon for Infantry in Her Majesty's Army.'[155] In fact later complaints were attributed to relaxed standards of workmanship, allowed to expedite supply to the East, and in particular reports of fouling leading to difficult loading were blamed on bad powder, bad oil and too fine cartridge paper.

Hardinge's open-mindedness on what was, in a sense, his contribution did not prevent him from concluding that the Enfield was 'the most deadly weapon ever invented and very effective both at a distance and close quarters'.[156] At 300 yards, a good marksman could reckon to hit a target six inches in radius.[157] At 800 yards the average absolute deviation from the point of mean impact was slightly over five feet, a margin of error certainly permissible against bodies of men. Its handiness and diminished recoil meant that while only 10 per cent qualified as first-class shots with the Minié, 33 per cent did with the Enfield. This factor compensated for the Minié's occasional marginally better performances at distances between 600 and 800 yards, although it is interesting to note that at 900 the Minié only scored 3.80 per cent hits against 10.66 per cent by the Enfield. At up to 500 yards, a moderate shot could reckon on 50 per cent accuracy. At extreme ranges, Douglas's objection to a high trajectory of fire was overcome: at 800 yards, the bullet fell six feet in forty yards: the shots that would thus pass over the front would still be effective to the rear, and those that pitched short would ricochet into the first rank.[158]

Apart from the manufacturing defects already noted, the pattern 1853 was extremely reliable. Fouling was cleared with each successive shot and continuous firing in trials was maintained for 150 to 200 rounds: in a test carried out a decade later, one Enfield fired 20 rounds a day until it had been discharged 16,000 times without cleaning.[159] With its gutta-percha stopper and Eley's patent caps (also lined with gutta-percha) Hardinge boasted that 'the Rifle loaded may remain a week in a pond and go off without fail'.[160] However, the first issues of the Enfield were not until late 1854, and even by August 1855, only twenty battalions in the Crimea were equipped with it, the remaining twenty-nine having Miniés.[161]

The improved finish and the rifled barrels of the Enfield and Minié demanded a standard of musketry higher than that of the Brown Bess, with its ill-fitting stock and simple fore-sight. Stories of soldiers in action loading two or three balls at one time or leaving the ramrod behind in the muzzle bear witness to a haste born of inadequate practice in the management of arms. Even in the Rifle Brigade, nominally specialists in the subject, it was reported in 1834, that 'From the habit of simply carrying their arms to and from parade without being accustomed to the use of them, the Men were so awkward, that a common field day or day's target practice, was seldom got over without the loss of a finger or two, if not some more serious accident.'[162]

A relatively minor controversy in musketry drill is illuminating on the standards of musketry training. It revolved around the use of the command 'Fire!', which appeared in the 1824 *Field Exercise* but not in the edition of 1833. The 1833 regulations allowed men to fire independently so that the whole line was not left simultaneously unloaded and exposed to a sudden cavalry attack: moreover it encouraged the soldiers to aim and fire only when their targets were covered, instead of straining their faculties listening for the order amidst the din of battle.[163] However, in volley-firing, the soldier had had time to find the true level as he waited for the order 'Fire!': with the single instruction 'Present!' he brought the musket to his shoulder with the finger already on the trigger and either risked premature discharge or fired hurriedly and consequently inaccurately. Trials in Prussia had marginally supported the 'Fire!' lobby, and experiments conducted by Chichester in Canada even more dramatically showed thirty-six hits and sixty-four misses if the order 'Present!' only was given, but eighty-nine hits and eleven misses when the final order was 'Fire!'[164] The discussion is interesting because it confirms that bad individual shooting stemmed from poor instruction.

As in so many other instances in the post-Waterloo army, the regulations for musketry instruction were specific and relevant, but largely ignored. The recruit, the 1834 *Manual and Platoon Exercises* said, must be taught to aim and fire with precision, so as '*to kill his enemy*' (italics in the original). 'The firelock', it went on, 'is placed in the soldier's hands for the destruction of his enemy; his own safety depends on his efficient use of it, and no degree of perfection he may have attained in the other parts of his drill can, upon service, remedy any want of proficiency in this.'[165] But the regulations were bound to be limited in their efficacy when the infantry were only allowed for practice twenty rounds of ball and sixty of blank per man per annum. The light infantry were marginally better off with fifty rounds of ball, and the rifle corps received ninety. If not claimed, 'the allowances will be considered as not required',[166] an instruction reflecting not only Treasury miserliness but also Horse Guards negligence, musketry not forming a subject of inspection. The Horse Guards, however, could legitimately claim that their hands were tied and that no uniform system of musketry instruction could be established when many stations did not have access to a range. In 1843 the Board of Ordnance confessed itself defeated by the obstacles put in its way when trying to secure ranges.[167] Even in Canada, the Rifle Brigade could only get a range of 200 yards with a single target, which promised 'slow work, for 500 men totally unpractised in the use of the rifle'.[168] The result was that

You are compelled to march a considerable distance to the sea side, where you fix your target on the beach on a ridge of shingle. The marker must be placed wide for fear of balls glancing, and the danger arising from hanging fire. Fatal accidents from these causes have arisen. The target furnished consists of an iron frame covered with the thinnest calico that ever came out of Manchester. At every supposed hit, the marker plods along the heavy shingle and ruddles or otherwise attempts to mark the shot holes. These multiply and run into each other; in vain he tries to discover and mark them. The calico becomes torn, balls fall short and force the stones against the calico, the marker cannot discover whether it is struck by them or the balls, all is perplexity, doubt and confusion.[169]

In such circumstances accusations of negligence against company officers are understandable if not forgivable. Often the whole company would march out to fire together. After waiting their turn to fire, and with delays for marking shots and correcting aim, the men would grow listless. The officers too would become bored, and, rather than think of a better method of instruction, would at last, 'merely to get rid of the ammunition', allow firing by sections, sub-divisions or companies.

Such practice . . . does men harm rather than good. A man fires expecting to miss, & of course does. If I had my chance, I would rather go into action with men who had never fired at all, than who had been accustomed to this sort of work: they at all events would have no reason to suppose that the muskets in their hands were only to make a noise.[170]

The author of these comments, Sir Charles Chichester, the commanding officer of the 81st, was one of a number who at least endeavoured to improve the musketry in their own regiments. Chichester trained his men in different types of fire – shooting from the hip, firing and then charging, target firing in heavy marching order, firing from behind a stockade. The result was that 'The fellows really think a musket is intended to hit what it is pointed at, that there is a method of handling it which will succeed, & try to do it.'[171]

Although other regiments took care to mark their shots or to present medals and prizes for shooting, there is little evidence that those which neglected such attentions to their musketry were put under any pressure to reform. A uniform course of ball practice was needed, not only so that an overall higher standard could be established but also so that regiments acting together in brigade would have been trained in similar types of fire.

'Ubique', alias Captain F. Robertson, writing in the *United Service*

*Gazette* in 1837, suggested that musketry should form a subject of inspection, that an approved course, both practical and theoretical, be drawn up, and that a central small arms school be established, whither one officer and two N.C.O.s per regiment should go for instruction. The recruit should not fire ball until he knew more of the construction of the musket, had been taught how to aim and to judge distance.[172] Major J. Bentham, of the 52nd, had a similar programme – ranges in every barrack, rewards for good shooting and a central school of instruction. Simple target shooting should be allied to more military types of fire: companies in extended order should fire as they retreated or advanced; file-firing, firing in close order or charging and firing should all be practised.[173] Other leading reformers who sounded off in the military press included Sir J. E. Alexander, Colonel J. Leach, and Colonel John Mitchell. The most telling contribution came from Captain Sorell of the 81st, in the *Aide-Mémoire to the Military Sciences*. He believed, in contradiction to the prevailing view, that most men could be brought up to 'a very fair average'; all that was required was an early start on a properly thought-out course. Each regiment should have an N.C.O. qualified to instruct in musketry, with an officer in overall charge, and training in theory and practice should proceed simultaneously.[174]

Official reaction was first stimulated by the Royal Engineers' report on experimental musket firing in 1846. Burgoyne submitted to Anglesey a memorandum urging that ranges be acquired and regular instruction in judging distances and target practice be provided. Both the Master General and the Commander-in-Chief agreed on the desirability of proper training with the Brown Bess, but the Commander-in-Chief, as ever, doubted whether financial support would be forthcoming. Only Fitzclarence, as commander at Portsmouth, implemented Burgoyne's suggestions.[175]

However, reports from France of the excellence of the *école du tir* at Vincennes kept the issue to the forefront of military debate. In 1850, another sapper, Sir Frederick Smith, reported to Fox Maule that a central school, as at Vincennes, and regimental instructors, as were attached to each regiment of *chasseurs à pied*, 'would be most beneficial, for then the reputation of one officer would be at stake in the result and it can hardly be doubted that a great improvement would take place'. He suggested periodical returns and musketry inspections. Not only would shooting improve, but also each man would learn the true worth and individual tendencies of his own arm.[176] Grey in particular conceived a tremendous, if exaggerated, respect for the training and musketry of the Chasseurs de Vincennes. But his own proposals seem to have been limited to the provision of more ranges.

The adoption of the Minié rifle made ranges only a part of the desiderata of musketry training. There was little point in possessing a reliable and accurate arm if its owner could not even judge distances in order to set the sight correctly. Consequently a Horse Guards circular of 6 March 1852 propagated rules for training with the Minié, giving instructions in the use of the sight, judging distance and in skirmishing and platoon firing. Ten target frames and ten traversing rests were issued to each regiment which had access to a range,[177] and, as the Minié was a rifle, the rifle allowance of ninety, not thirty, rounds of ball was granted.[178]

These provisions might have had only limited effect had not the first faltering steps towards a centralised system of musketry been taken in February 1852. All regiments due to receive the Minié were instructed to send one officer and two N.C.O.s to Woolwich to undergo a course in the mechanism and use of the rifle, with a view to returning then to their own corps to impart their new-found wisdom. Independent firing, file-firing, skirmishing while retreating in extended order, volleys in sections, half-companies and companies, all at ranges from 200 to 900 yards, were taught.[179]

Thus far had matters already proceeded when Hardinge arrived at the Horse Guards. It is not surprising that a man so concerned for the gradual improvement of the soldier's firearm should also cherish a deep-seated belief in the value of musketry training. Hardinge related efficiency for field service directly to rifle practice and refused to allow under-instructed men to proceed to the Crimea.

He therefore planned to put the Woolwich course on a more permanent footing so as to establish a uniform system in the army. Colonel Estcourt was despatched to Belgium, Prussia and France to examine their methods of training, and in February 1853 a range of 1,400 yards was secured at Hythe within half a mile of the barracks. In May, the School of Musketry was opened,[180] and certain regiments were instructed to send a sergeant and nine other ranks to form 'an initiatory school of firing on which to base the future system to be pursued'.[181] The pupils were kept on after their instruction was complete, in order both to carry on experiments with the pattern 1853 and also to evolve a new platoon exercise and a finalised musketry course. The new platoon exercise facilitated loading and rendered the soldier's position 'more easy and natural'. The motions of the front and rear ranks ceased to be distinct, a better balanced stance, with the feet apart, was adopted, and when kneeling the rear leg was tucked up underneath the body, thus giving the firer a steadier rest and lower profile.[182] The new book, published in June 1854, insisted that whenever a soldier brought his firelock to the present he should take deliberate aim at a specific object.

Nonetheless, the main achievement of the opening year of the Hythe School was the army's first official manual on the subject, the *Instruction of Musketry*, published in February 1854. Each regiment was to have two Hythe graduates, an officer and an N.C.O., as instructors of musketry. Company officers and N.C.O.s were still to bear responsibility for musketry training, but the N.C.O. instructor had to attend all target practice and the officer had to lecture on the theory. Practical training fell into two classes: target and position drill, judging distances, the manufacture of cartridges and the cleaning of the rifle formed one part; in the other came target practice, including firing singly, file-firing, and firing in extended order, with skirmishing as an optional extra. The recruit was allowed an extra twenty rounds a year to be fired from a rest, giving him 110 rounds in all. A precise scale of practices and drills to be carried out within a year was established, with the number of rounds to be expended in each specified. Each shot fired was to be registered and returns for every practice were to be handed in.

This, the Hythe system, was not operational at the start of the Crimean War. The experimental practice required to draw up such a comprehensive syllabus meant that the first detachments proper did not arrive at the School of Musketry until April 1854. Because it was conceived as a three-month course with a maximum of twenty officers in attendance, its influence could only gradually percolate through the army. Furthermore instructors for the school itself could hardly be expected to emerge ready-formed: Major John Clark-Kennedy was sent on a tour of the manufacturing establishments at Waltham, Purfleet and Enfield, and then raided the Tower Armouries for duplicate weapons for the Hythe lecture room, before taking up his appointment as instructor at the School.[183] Therefore, to help train the expeditionary army as rapidly as possible, Captain Lane Fox was sent from Hythe in March 1854 to instruct at a similar school in Malta, a staging-post for the Bosporus.[184]

To complement the central organisation, it became essential to fill a longer-standing and more grievous need – safe ranges close to all the main military stations. Three new ranges were established in 1850–1,[185] but beyond this, if we accept the evidence of the military press, provision within the United Kingdom was minimal indeed. In May 1853, District Commanders were to inform the Commander-in-Chief what difficulties, if any, lay in the path of regular musketry practice and instruction. The replies indicated the scope of the problem: cattle grazing interfered with firing at Winchester, and high tide ended proceedings at Newport and Dumbarton. Several ranges were only four hundred yards long, and thus perfectly adequate for the Brown Bess, but too limited for the Minié. Swansea had excellent facilities, but no troops were stationed there.[186]

Nonetheless in June, the Horse Guards were enabled to draw up a list of ranges eligible for use with the new rifle, giving twenty-two in Britain and a further thirty-eight in Ireland. Almost all of these, however, were on the coast and exposed to cross winds or tides. Thus most inland stations were neglected, and in England no suitable ranges were to be found between the south coast and Liverpool.[187] The 2nd Battalion of the Grenadier Guards, for example, had gone five years without ball practice, until in August 1853 the Guards secured a range in London, at Kilburn.[188] During the second half of 1854, a further twelve ranges at least were set up or improved, although the tendency to site them on the coast still continued.[189] Inland, intensive cultivation, populous towns or regard for the inalienability of private property, with their consequent effects on land prices, precluded rapid expansion.

Hardinge, however, would not be defeated by these necessary limitations. District returns of rifle practice were required every week from April 1854,[190] and from the following month the first forty days of recruits' instruction were almost entirely taken up by a progressive course in the rifle and its use.[191] Writing in November 1853, Hardinge confidently predicted that 'in a couple of years, our shooting will exceed that of the C[hasseurs] de V[incennes] or all Europe'.[192] Accounts of Alma and Inkerman confirm that to a certain extent the Commander-in-Chief's hopes were realised within a year, thanks not only to the native virtues of the Minié but also to the extensive practice which Hardinge ensured that all troops for the East underwent.

As significant as the range and dexterity achieved by the British soldier under the Hythe system was the emphasis on types of fire other than that of the close-order line. Seaton correctly foresaw in 1853 that the instruction given in an *école du tir* enabled all infantry to be 'employed in extended manoeuvres advantageously, in covering deployments, in skirmishing in woods, in harassing columns of attack, and in keeping down the fire of forts and batteries'.[193] The tendency for all line infantry to act as light infantry was confirmed by the wholesale adoption of the rifle. As the *Military Review* acknowledged, progress fell into three distinct heads, 'The greatest possible development of the Fire-fight', 'The adequate remodelling and simplification of the Tactical formations', and 'The abrogation of the impractical and complicated partition of armies into Troops of the so-called different arms' (by which it meant the division into 'light' and 'heavy').[194] Hardinge, more succinctly, but for a Peninsular veteran more pregnantly, said that as soon as the rifle is issued, 'There will in fact be no distinction, but that of the Colour of the Dress, between Rifle Corps & Infantry.'[195]

For, quite apart from the weapon itself, the universal adoption of the rifle in place of the musket worked, in the opinion of a Fitzclarence protégé, Colonel Maitland, 'a complete revolution in the art of war'.[196] The defence was seen to have regained the initiative. In the three minutes the cavalry took to cover 800 yards, the Enfield could fire twelve rounds at them without reply.[197] Field artillery were reckoned to be ineffective at much beyond 600 yards: gun crews would be exposed to infantry fire but could not fire back. The close column of attack had clearly had its day. William Napier, in a woeful mood, painted his view of the consequences: the Minié 'must paralize the action of cavalry against infantry and of artillery within that range. No close reconnoitring can take place as the generals would inevitably be killed, & hence battles will be more confusedly arranged more indecisive & more bloody.'[198]

His conclusions need not have been so irredeemably negative, since the rifle itself presented a partial answer to the potential impasse. The increased power could be applied not only to the defence but also, in the shape of light infantry, to the attack. Colonel F. R. Chesney, quoting Captain Wittich of Prussia, saw the solution in the best organisation of the inevitably increased forces of light infantry upon which an army would rely. Small columns, four ranks deep, divided into platoons or half-platoons were envisaged. Firing would be by alternate ranks, but, once the attack was under way, no delay could be brooked because of the larger distance to be traversed under fire.[199] Colonel Gawler, in an updated edition of his 1837 work on skirmishing, stressed that the company, not the battalion, must be the main tactical unit, and, to retain cohesion, movements should always be based on the centre. Skirmishers should learn to drop on the knee when halted, and double when moved. Firing and advancing should be by alternate ranks.[200]

Such a combination of fire and movement should in fact have given heart to the Napier brothers. The emphasis on these skills and the difficulties of maintaining a supple formation under fire put a premium on higher training and in turn should have aided, rather than harmed, Britain's long-service professional army. But in 1854 the British authorities were only just beginning to digest the killing power of the Minié, without fully appreciating its tactical consequences. Hardinge realised 'that all defensive positions are more easily defended now that a well trained Inf. can hit its assailants at 800 yards and under cover whilst the advancing force may be murderously exposed'.[201] Seaton, equally cautiously, wrote in his instructions for Chobham camp that the opportunity presented itself to decide 'whether every Regiment of Infantry armed with the improved Rifle Musket ... might not, with the greatest advantage to our Army in the field, be employed, on all occasions of Rifle Service, as Rifle Corps'.[202]

Their thinking had not yet progressed sufficiently to supplant the 1833 *Field Exercise*. It was the latter that dictated infantry tactics at the Alma and at Inkerman. Equally, it was the traditional light troops who skirmished on those occasions. Indeed the Russians commented on the slow, steady gait of the British line, when compared with the alacrity of the light infantry.[203] Current thinking perhaps vindicates Sir Howard Douglas in reckoning that 400 yards is the maximum range of normal infantry fire: now, as in the Crimea, the sniper is a specially selected man. Immediately, however, by giving the infantry arms and training in which to have confidence, Hardinge improved the musketry of the line dramatically. At Vittoria, one bullet in, at the best, 459 took effect; at the Alma and Inkerman, one in sixteen.[204]

On the European battlefield, it was to take technology only a few more years to apply the rifled breech-loading principle to artillery, and thus to remove the initiative from the infantry. However, in the colonies, the difficulties of transporting large quantities of artillery and the unlikelihood of its possession by the enemy meant that the revolution in firepower gave the British army a qualitative superiority in armament over a much longer period and against every enemy except the Boers. The tactical assets it thus conferred compensated in considerable measure for strategic inferiority in terms of intelligence, immunity to disease, transport difficulties and unfamiliar terrain. Furthermore, although applicable to attack in open order, disciplined fire's greatest bonus was conferred on the defence. By employing a strategic offensive with a tactical defensive, the British army could tempt large native forces into the range of its rifles. Thus it could achieve much with small bodies of men. Thus too its tactics conformed not only to economic and imperial necessity but also, in some respects, to Wellingtonian precept.

# 3

# Cavalry: the blighted flower

## (i) Cavalry on the battlefield

William Napier's glorification of the British infantry had a contrasting aspect. In the history of the Peninsular War, 'the cavalry alone has been excepted from the boon of unmixed eulogium, and has even been censured by many as not having fulfilled its allotted share in the great contest'.[1] Despite its highly distinguished record in the Low Countries in the 1790s and its crucial part in covering Moore's retreat to Corunna, Wellington had rarely employed his *arme blanche*. Apart from advance guards and flanking cover, the Duke allowed it to pursue beaten troops and, in conjunction with infantry, to attack small groups of enemy cavalry. Of course his supply of horsemen was slender and consequently had to be nurtured; of course too his tactics tended always to the defensive, but his own assessment of it was never flattering – inferior to the French and 'the most delicate arm that we possess'.[2]

Wellington's derogation of the British cavalry, supported by the judgement of William Napier and the opinion of the enemy, was sufficient to give that arm a very humble view of itself for the next half-century. Brotherton, a regimental cavalry commander in the Peninsula and Inspector General of Cavalry in the second half of the 1840s, agreed with Napier's conclusions on the subject of the cavalry in the Peninsula.[3] Except for Combermere at Bhurtpore, virtually every officer who exercised an independent field command from Waterloo to the Crimea was an infantryman trained in the Peninsular school. If Brotherton is a fair gauge, there existed little pressure for these officers to revise the Wellingtonian view of the cavalry's limited potential.

Inherently national factors therefore made it improbable that British cavalry would be employed in the grand manner. But in addition all cavalry were challenged by the example of the British infantry squares at Waterloo. The Frenchman F. de Brack, in his *Advance Posts of Light Cavalry*, instructed his budding Murats that a solid square was unbreakable, and that they should therefore wheel round the square and call out to the enemy that they were prisoners.[4] Major N. Ludlow Beamish in his edition of Bismark's *Lectures on the Tactics of Cavalry*, which was to be the only major work in English on the subject until 1853, wrote, 'The *infantry* is the base and foundation of all armies; it is the commanding power, which may be called the *soul* of war; ... cavalry is an accessory.' The conclusion to such thinking was a truly Wellingtonian precept: 'If infantry is supported by artillery, and formed in squares, it is almost impossible for cavalry to break them.'[5]

Logically Beamish seemed to be correct. Over the final 400 yards of a charge, cavalry would probably receive three volleys, and, even allowing for the inaccuracy of the smoothbore musket, the final rounds ought to have left a rampart of dying men and horses between the square and its attackers. Even at Salamanca, where the French infantry were not in square, the successful British cavalry had suffered one-third losses. Should the cavalry actually close to the muzzles of the square, the determination of the riders would be countered by the horses' natural desire to avoid trampling people. At Barquilla, the 14th Light Dragoons charged so well that the French bayonets were broken and musket barrels bore sword cuts, but the square still stood intact.[6]

Nonetheless instances could be quoted of infantry squares being broken by cavalry charges. As even Charles Napier, a protagonist of the foot-soldier, cautiously confessed: 'Infantry in *squares* and behind *stone walls* seem to be the only things that Cavalry cannot overthrow, and even these demand *certain conditions*.'[7] The principal condition for the infantry was, as Colonel Gawler put it, 'a well impressed knowledge of their own strength'.[8] Infantry manuals tried to impart this confidence by stressing repeatedly that well-formed squares could not be broken by cavalry. The dominance of the infantry in British military thought and the comparative poverty of cavalry treatises gave force to an assertion that gained strength as much by repetition as by its own validity. Major General F. Ponsonby, who as a cavalry commander needed to believe in the abilities of his own arm if he were to gain success, confessed to Sir Andrew Barnard, 'No body of cavalry can hurt a well formed battalion of infantry, if conducting itself steadily and reserving its fire till the cavalry arrives at about 100 yards from the column. You recollect the story of the keeper who used to put his head in the Lions mouth and ask if the Lion shook his tail, so it is with us, if in charging infantry we see the shake, we know we have them if not we know we shall be b——— f———.'[9]

In Europe this self-defeatism on the part of cavalry had not reached such epidemic proportions. Methods of restoring the initiative to the cavalry, of emphasising that it was the decisive arm of attack, and that the value of morale was as important to it as to the infantry, were therefore derived from France, Prussia and Austria rather than home-bred.

The first such attempt was the aping, even down to the details of costume, of the Polish Lancers in Napoleon's service. In 1816, four light dragoon regiments were equipped with the lance. It was claimed to be 'the sole weapon with which Cavalry can hope for success against the fence of bayonets presented by a square of infantry'.[10] Marmont argued that 'If . . . Cavalry is armed with a range of lances, which project beyond the horses four feet, nothing can be opposed with success to its efforts.'[11]

The British lance was made of ash, was just over nine feet long, and therefore projected only a little over four feet from the rider rather than from the horse. Prolonged trials in 1837–40 with bamboo lances, whose lightness would permit greater length, were not deemed successful,[12] and so the alleged reason for using the lance at all was rendered null. Bismark, when he spoke of the lance making 'the cavalry again capable of attacking infantry with success',[13] was talking of a weapon eleven feet long. Nor did the weight of the ash compensate for its shortness, since against the Sikhs the 9th Lancers often failed to pierce shields, armour and even wadded clothes.[14] Accounts of lancers reaching over into squares do not seem to have survived, and the value of the lance is more often reported to have been in stretching down to tribesmen who lay on the ground as the cavalry passed. Technically even this could be achieved by a well-balanced, long-bladed thrusting sword, like the 1908 cavalry pattern.

The other argument in favour of the lance was its moral effect, its formidable appearance being symbolic of the medieval horseman. But, while it thus stimulated the prose of despatch-writers, it actually obscured the responsibility for a successful charge. After all, the lance depended for its efficacy on the momentum of the horse. Furthermore, apart from a minor but victorious action at Bhurtpore, the only real test of the weapon against well-formed infantry was the charge of the 16th Lancers at Aliwal in 1846. An N.C.O. who rode in the charge asserted that it was the lance alone that broke the Sikh square,[15] but the latest historian of the British cavalry doubts this, and at least one contemporary account describes a horse as breaking the front rank. In parentheses, it is noteworthy that the 16th's casualties on this occasion, 141, bear testimony to the doubts about cavalry charging infantry at all.[16]

Although lancers in the British service were nominally light cavalry, the charge at Aliwal was essentially, in terms of European warfare, a heavy cavalry action. Sir Samford Whittingham contended that in India the roles of the French cuirassiers in the charge and the pursuit were more naturally fulfilled by lancers in line, 'able bodied men mounted on powerful Horses'.[17] Beamish agreed that the lance was a heavy cavalry weapon.[18]

Fig. 9 *The Charge of the 16th (Queen's Own) Lancers at the Battle of Aliwal, January 28th 1846.* Engraving by J. Harris after H. Martens (National Army Museum)

Russia gave the lance to the front ranks of all her cuirassier regiments, the argument being that the lance would break the infantry and the second rank would be free to use their swords in the ensuing mêlée. Sir Hussey Vivian made a similar recommendation to de Lacy Evans for the cavalry which he took with him to Spain in the British Auxiliary Legion.[19] Rumours that heavy cavalry in Britain were due to receive lances, at least experimentally, abounded, and it seems probable that trials were conducted at the Cavalry Depot. The Duke of Cambridge, who cited the Austrian success with the lance in 1848,[20] urged that the lancers should recruit bigger men: 'The Lancer has more of the character of the Heavy Dragoon about him and as such should be an inch taller ... A lance in the hands of a powerful man is a magnificent weapon, perhaps the finest in the world, but in the hands of a slight man it is more than useless.'[21]

The Duke, as Inspector General of Cavalry at the outbreak of the Crimean War, and a former 17th Lancer himself, was thinking of a force fighting a European war. However, the British army was more likely to be committed to colonial operations. More relevant was the view of Captain John Kinloch, that the lance was better adapted for irregular troops.[22] The fame of the Cossacks suggested that, despite its size and inconvenience, it was compatible with scouting and reconnaissance. The Cossacks in the Caucasus were conducting operations similar to those British cavalry might undertake. But, unlike the British, the Russian cavalry was a massive force capable of bearing specialised subdivisions. The lance was useless in a mêlée with other cavalry, and even infantry, it was argued, had to be broken before the lance could be brought to bear. The British instructions for the lance exercise, that the opponent should be kept to the front, to the right and at a distance (so as to gain the maximum advantage from the weapon's length) point to its inflexibility in action.[23]

Therefore, whether it was seen as the heavy cavalry's answer to infantry firepower or as the complement to the light cavalry's roaming brief, the lance was probably ineffective and certainly demanded excessive specialisation in an army where only one cavalry regiment was likely to be in a particular quarter of the globe at any one time. *Pace* the 16th Lancers at Aliwal, Captain Humbley of the 9th Lancers even questioned the value of the lance against the Sikhs, an army trained on European lines.[24] The 12th Lancers, while at the Cape in 1853, were instructed to leave their lances in store and use carbines.[25] Although this made nonsense of their training up to that point, that this was the right decision was shown by the ineffectiveness of those who did not conform to the order. Sir Harry Smith was planning to use them in the more open country across the Kei river, and therefore two-thirds of the regiment went on carrying lances.[26] Consequently infantry officers complained that their cavalry was useless, and at the battle of Berea Sir George Cathcart found the 12th 'too heavy and unwieldy'.[27]

Paradoxically one of the most trenchant critics of the lance was an officer whose experience, although it embraced service in India, was fundamentally continental. Captain L. E. Nolan of the 15th Hussars averred that, whether the lance was used in single encounters or by heavy cavalry, 'The moment the lancer pulls up and the impulsive power is stopped, that instant the power of the weapon is gone.' Cavalrymen with experience in the Sikh Wars had assured him of their condemnation of lances, and 'It is well known that, in battle, lancers generally throw them away, and take to their swords.'[28]

Certainly the sword was handier, easier to manage in close combat and less inconvenient on outpost. Nonetheless its destructive powers as a weapon of war were strangely limited. At Mudki in 1845 an officer of the 3rd Light Dragoons had difficulty cutting through the cotton clothing of the Sikhs[29] and at the battle of the Gwanga the Kaffirs struck by sword-blades showed a remarkable capacity for quick recovery.[30]

Much of the trouble stemmed from the most obvious of faults. The Sikhs

Fig. 10 *Charge of the 3rd (King's Own) Light Dragoons at the Battle of Ferozshuhur, December 21st 1845*: light cavalry acting as heavy cavalry, and giving both cut and thrust with their swords. Engraving by J. Harris after H. Martens (National Army Museum)

kept their curved blades sharp in wooden scabbards. British blades were blunted by the constant banging and rattling in steel scabbards, and the same officer who experienced difficulty at Mudki found that a little sharpening rendered his sword effective at Sobraon. A second problem was the preference for cutting rather than thrusting; John Jacob, in common with other cavalrymen, found it difficult to withdraw a thrusted sword when mounted on a fast-moving horse.[31] A participant in the battles with the Sikhs said that in close combat, few thought to give point, and instead cut in the Indian fashion.[32] Indeed in a mêlée it could be hard to find the space or opportunity to thrust. The Sikhs, however, rode with a shorter stirrup and tighter rein than the British were trained to do, and thus in a mêlée were in a much stronger and more manoeuvrable position to deliver a blow.

The British sword, the 1821 light cavalry pattern, was defined as a cut-and-thrust weapon. In fact the point of balance was very near the hilt, the strongest part of the blade being here so that the swordsman could parry effectively. The remainder of the slightly curved blade was so light that those first issued tended to break. Following a report of 1828, swords made after 1836 had straighter blades.[33] The purpose of this alteration was to underline that in attack, the giving of point, not cutting, was to be the prime motion. But it was a fallacy that the same sword could efficiently both cut and thrust; the former needed a shorter, sharp, curved and heavy blade, the latter a long, straight one. In leaving scope for cutting at all, the Horse Guards had weakened the weapon.

Its successor, the 1853 pattern, was merely an improvement of the same basic design. Issued initially to light, but eventually to both heavy and light cavalry, its principal feature was the tang of the blade, which was let into the grip, and thus gave a much securer join between blade and hilt than its predecessor had boasted. The blade itself remained the same as on the 1821 pattern but the virtually cylindrical grip probably made it difficult to cut with in the heat of action.

The performance of the 1853 pattern in the Crimea is hard to gauge, partly because it is not known how widely it was issued. The 11th Hussars, before sailing, tested one of the new swords and found it so soft that it bent 'like an iron hoop', and the 2nd Dragoons complained that they 'bend and will not rebound'. The proofs to which these two regiments subjected their weapons were considered excessive, and, since the Greys did not receive their replacement swords until at least November, the commanding officer's complaint that, in the regiment's charge at Balaclava 'when our men made a thrust, with the sword, they all bent and would not go into a man's body', was partly his own fault.[34] By contrast Evelyn Wood reported that in some cases the skulls of the Russians were cleft to the chin, an indication of what, despite its constructional deficiencies, the edge of the cut-and-thrust blade could achieve if properly sharpened.[35]

The thickness of the Russian coats apparently compelled recourse to the cut rather than the point, but on the same day Colonel Douglas reflected the new orthodoxy when he instructed the 11th Hussars 'don't cut, but give them the point'.[36] Wounds inflicted by the point were more fatal, and Colonel Luard of the 16th Lancers lobbied for a sword so straight and long – four feet – that the trooper could only give point.[37] Charles Napier[38] and Major Beamish[39] both saw the ideal cavalry weapon as a long straight blade, the latter favouring thirty-six or thirty-eight inches as the length. Furthermore the ease of thrusting when mounted on a charging horse encouraged the cavalry to shed their penchant for cutting. Fighting Afghans in the Tezeen valley, the redoubtable Captain Unett found that, with his horse bearing him (he thought) at about twenty miles an hour, his sword passed through a tribesman 'without the slightest exertion'.[40] 'Let the horse out to full speed . . . do not in any way interfere with his mouth . . . and let the horse carry the point home', Nolan urged.[41] The superintendent of sword exercise correctly observed that, with the horse in motion, an extended arm and a good direction were sufficient.[42] As with the lance, the responsibility for success lay less with the weapon than in the determination of the horse and rider.

The only other auxiliary available to charging cavalry was the pistol. Even more than was the case with infantry firing as they advanced, the regular use of the pistol broke the rhythm of the onset. Marlborough had dispensed with the Cromwellian practice of pistol-firing, encouraging cavalry to rely on the sword alone. The result was that, in the 1833 regulations, the pistol was defined as merely a useful standby at close quarters if the sword or sword-arm were disabled. The fire of the pistol was uncertain, short and seldom efficacious, regarded as an encumbrance by those who carried it. In 1828, a Board of General Officers, with Vivian as president, reported that 'It is the opinion of every Officer Commanding a Cavalry Regiment, and of, it may be added, every old soldier who has had experience of European Warfare, [that the pistol] is seldom, if ever, found of any use.'[43] Consequently in 1839 pistols for Light Dragoons and Hussars were abolished, except for officers, sergeant majors and trumpeters, who were not, however, issued with practice ammunition unless it was specially requested.[44] The lancers, who were already too encumbered to carry a carbine, persevered somewhat unavailingly with the pistol.

Nor did the appearance of the revolver usher in a new era for the cavalry charge. When Colt's model was displayed at the Great Exhibition in 1851, it was accompanied by a reputation won for it by the United States Dragoons in the Mexican and Indian wars. Four hundred and fifty Colts were instantly despatched to the Cape,[45] an American salesman following hard on their heels, and in six months revolvers had become 'much the fashion out here'.[46]

Even if he had been first off the mark, Colt was not destined to sweep the market unopposed. Although his revolver had six chambers, it still weighed four pounds despite its comparatively small calibre (60 balls to the pound). It required to be recocked after every shot and thus inconvenienced the mounted man who was left without a spare hand for his reins. At a Board of Ordnance trial in 1851, it had taken fifty-eight seconds to load, and a further eleven seconds to fire, only hitting the target at fifty yards when used in conjunction with a rest. At one stage it had five and then seven consecutive misfires. Its rival, the Deane and Adams, had only five chambers, but this enabled a larger calibre (32 balls to the pound) and a weight of only 2 lb 14 oz. Not surprisingly it was more accurate, could be used much more conveniently with one hand, was loaded and fired in half the time, and never misfired. Both claimed a much better performance than a rifle at up to 100 yards.[47] The bias of the Ordnance Select

Committee in favour of the Deane and Adams was therefore prompted by more than patriotic sentiment.[48]

However, despite its other deficiencies, the Colt had one great military virtue: it was machine-made, its parts were therefore all interchangeable and repairs were cheap. The London factory could meet the large and urgent orders which the Crimean War generated. From the cavalry viewpoint, the powder of the Colt was not shaken out by riding, whereas the ball tended to loosen in the Deane and Adams. Consequently it was the Colt that Colonel Chalmer recommended for service in March 1854.[49] It was issued in August to the Mounted Staff Corps,[50] and the following year the Land Transport Corps received it;[51] 1,500 were allocated to the infantry in August 1855.[52]

Nevertheless mistrust of the new weapon still predominated. Many officers equipped themselves with revolvers on going east, but the cavalry members of a Board of Officers convened in the Crimea in January 1855 rejected as unsuitable for mounted men 5,000 that had been sent out, claiming in particular that they were constantly out of repair and liable to go off accidentally.[53] In the summer of 1855, when the supply of pistols was exhausted, Colts were delivered to the lancers, but it was made clear that this was not to be a general issue.[54] C. W. Pasley was probably correct when in 1851 he anticipated that the value of the revolver would be confined to the charge or mêlée, and yet in such circumstances, as Cardigan himself found out when beyond the Russian guns at Balaclava, it was hard to remember that the weapon was there, let alone contrive the opportunity to return the sword to its scabbard and draw the revolver from its holster. Whatever its achievements under trial conditions, the effective range of a revolver used by a mounted man was calculated at little more than ten yards.[55] A vast improvement on the pistol it may have been, but the revolver remained a weapon of personal defence, and was not a new element on the battlefield or the panacea for the cavalryman's problems.

The failure of technology to provide a new armament suitable for mounted men forced the theorists to revert to first principles. As the successful use of the lance and sword demonstrated, the cavalryman's most powerful auxiliary was not a weapon but his horse. 'It must be a general rule', the 1844 *Cavalry Regulations* instructed, 'never to break or depress the noble and generous spirit of the animal.'[56] The horse's courage was more than an adjunct to the delivery of the *arme blanche*; it was, according to Marmont, the cavalryman's 'true offensive arm, which being killed, falls and makes a breach [in the infantry], which affords some of his neighbours the means of penetrating into the square'.[57] Aliwal seems to have proved his point, and at Willems in 1794 and Garcia Hernandez in

1812 it had certainly been the weight of a horse that had created an opening.

In Britain, the case for unfettered cavalry was opened by Colonel Mitchell. Infantry, he stated, were quite simply powerless against well-directed cavalry. If cavalry charged infantry, either they must be shot (but few are seen to fall), or be impaled on the bayonets but nonetheless break the square, or veer off to the side. Infantry's vulnerability in line was commonly accepted, but in theory a square four ranks deep presented no more formidable a target. The difference was that the flanks of an extended line could be remote from the direction of the charge and therefore force it to ride straight, whereas the intervals between squares encouraged it to break off round the sides. The tendency for this to happen was reinforced not least because on field days it was what horses were trained to do. Only a quarter of a square's firepower could be used against the cavalry, and that was normally reserved till they were within fifty yards, when the horses were already into the final gallop and, even if dead, would not collapse until on the square. Mitchell computed that of eight regular charges made by British cavalry against French infantry in the Revolutionary and Napoleonic Wars, two failed because of outside circumstances and one on its own demerits, while the remaining five were successful.[58]

Mitchell's argument was taken up by those who had experience of continental warfare. In Europe, the tradition of Frederick's cavalry was strong and conscript infantry were likely to prove less resilient than the British foot. An anonymous visitor to Russia came back arguing that 'Not once in a thousand times would every horse be hit; but even if it were so, it takes at least an average of five shots to bring down an animal like a horse, so that he shall drop on the spot on which he is hit.'[59] A former British cavalry N.C.O. who had served in a foreign army, probably in Spain, put his faith so squarely in the horse that he instructed the front rank to sling their lances, blindfold their horses and ride with reins in both hands. The moment they saw the enemy about to fire, they should charge immediately, the second rank following in before the infantry had time to reload.[60]

Nolan, reflecting his years in the Austrian army, could not agree with Mitchell that cavalry alone would always break a square, but only that they might frequently do so. If they did not, the threat of their use impelled the infantry into close formations peculiarly vulnerable to artillery or – as the Afghan jezails had shown – to long-range rifle fire. Therefore horse artillery used in conjunction with the cavalry, to break up the infantry out of range of its muskets, would always ensure the success of the charge. Two sides of the square should be hit simultaneously, with a third attack following in at the corner. The 1848 Hungarian war proved that neither modern artillery nor infantry firepower was enough to check cavalry who did their duty. Like Mitchell, he reminded cavalrymen of their successes: Frauenstadt, Hohenfriedberg, Zorndorf, Avesne-le-Sec, Villiers-en-Couche, Cateau Cambresis, Edesheim, Kaiserlautern, Nordlingen, Austerlitz, Auerstadt, Waltersdorf, Pultusk, Garcia Hernandez, the Goerde, Haynau. Even in the 1850s these names seemed to promise success for the *arme blanche*, and for the infantry there was scant consolation: 'all regulations are silent as to what is to be done when [the cavalry] do not turn, after the last shot has been fired and the horses are on the bayonets'.[61]

In the ultimate analysis success depended on the moral courage of the riders, and the vigorous example of their leaders. The sudden onset of a determined cavalry must give them the balance in boldness. Timing, an eye for country and rapid decision were the key factors: 'Of so much importance is the right moment in the handling of this arm', Major Jones of the 11th Hussars wrote, that a cavalry commander 'ought to be a real soldier, firm and ready to seize all opportunities, without waiting instructions.'[62]

Rid of Peninsular preconceptions, the rising generation of officers was clear what the requisites of cavalry leadership were. In criticism of Gough's handling of the cavalry at Chillianwallah, Havelock, although a humble infantryman, could see that 'Cavalry is essentially an attacking, not a defensive force ... The cavalry soldier that deliberates is lost.'[63] Nolan, in discussing the same misdirected use of the 3rd Light Dragoons, bewailed the fact that 'Modern tactics hold the cavalry in leading-strings. The system is a timid one, and made up of "ifs" and "buts", *words* which ought to be unknown to cavalry soldiers.'[64]

The opportunities to prove the worth of cavalry came infrequently and most often to young officers whose experience was negligible. How were they to grasp the timing, to act spontaneously in the field? A reading of Bismark suggests that inborn characteristics were the only resource: 'the secret of producing great effects by cavalry lies in the *personal qualities* of the leader'.[65] Nolan admitted that a 'system' for cavalry was a virtual impossibility.

> With the cavalry officer almost everything depends on the clearness of his *coup-d'oeil*, and the felicity with which he seizes the happy moment of action, and, when once action is determined upon, the rapidity with which his intentions are carried into effect. There is little time for thought, *none* for hesitation; and once the movement is commenced, its successful accomplishment is the only thought allowed to pass through the mind of the commander.[66]

The hunting-field seemed to be the only place where such attributes could be acquired. 'The duties of the cavalry soldier in the field', Colonel Luard of the 16th Lancers wrote, 'are occasionally more like the work of the fox-hunter than is generally supposed.'[67]

But for the Peninsular school, such arguments, such nebulous definitions smacked of special pleading designed to prolong the idleness of the rich, young cavalry subaltern. Because he was more likely to be operating independently than his harder-working pedestrian friend, he should apply himself all the more sedulously to the theory of war. 'I cannot help saying', wrote Charles Napier in 1840, 'that cavalry officers are particularly called upon to study war. Those of a low rank are often obliged to rely upon themselves: more often than are those of the infantry.' He spoke of cavalry being capable of anything if well commanded and if their 'prodigious force' was effectively applied. They could check infantry from changing their position, they could turn the enemy's flanks or protect their own, seize a post or an artillery battery, attack other cavalry, cover retreats, reconnoitre, harry convoys, and force infantry into squares so that they would be exposed to artillery fire.

Rarely can cavalry be called upon to charge squares of infantry, still more rarely will such rashness succeed, but a cavalry chief who has a quick eye, an intrepid spirit and his squadrons well in hand may seize opportunities of both a moral and physical nature such as surprise, panic, false movements; and such weaknesses. These moments are very precious in battle. They may exist for a moment; then pass away at the instant; and are lost for ever![68]

Although going considerably further towards the cavalryman's position than many of his contemporaries, Napier did not put cavalry on the pedestal that Nolan coveted for it. The conqueror of Scinde still saw cavalry as an ancillary arm, which might occasionally by a freak combination of circumstances decide the battle but which would probably spend its time in reserve waiting for unknown dangers. However, between Nolanesque flights of fancy and Napierian reservations, there existed a middle way. Despite the doubts, a rough system was hewn from the generalisations, and on the whole it not only improved the cavalry's efficiency but also restored to it some of its self-respect from the nadir of 1815.

At its lowest point, the self-defeatism in the years immediately after the peace even attempted to destroy the union between horse and man that was the mainspring of a cavalryman's effectiveness in combat. Lieutenant-Colonel J. G. Peters, in a manual on equitation first published in 1817, advocated that the soldier's seat be as near perpendicular as possible and that the stirrups' primary purpose be confined to mounting and dismounting. These instructions were embodied in the official manual of 1819 and, although not as extreme as the Prussian regulations which required an absolutely straight leg, they were still the direct cause of ruptures and other discomforts. The seat was held principally by balance, so that in riding long distances the chances of the mount's developing a sore back were

Fig. 11 a and b *The Military Seat as it is* and *The Military Seat as it ought to be*. From L. E. Nolan, *Cavalry: its History and Tactics* (1853) (by permission of the Syndics of Cambridge University Library)

minimal. Moreover, in action the British trooper would not be incapacitated by having his stirrup leather cut. However, in a mêlée, his control of the horse was reduced, he could not turn in a short space, and he could not rise in his stirrups to deliver a downward cut – as good a reason as any other for giving point. The 1819 regulations were gradually modified. By 1835, the cavalryman was allowed to ride on his bottom and not on the fork, and the stirrups were marginally shorter.[69] In 1851, Corporal Major Turner, the Assistant Riding Master of the Blues, laid down that the knee should be bent, with the foot exactly under the knee; in this position the stirrups should touch the rider's ankle before he inserted his feet.[70] Nolan advocated a hunting-seat, but this made little provision for successive days in the saddle.

As though Britons were not themselves sufficiently practised in the equestrian arts, Prussian influences even embraced the country's system of cavalry drill. Again the original author, as with the infantry, was Dundas. Too many principles applicable to infantry had found their way into his book. In particular, the line was controlled from the flank, which meant that the force of example and, more importantly, the control of a charge's direction, were put in jeopardy. Secondly, all the complicated additional measures designed to prevent the infantry inverting (reversing the order of the troops by turning about) were repeated in those of the cavalry: movements by threes and not by divisions ensured the minimum of flexibility. No fundamental revision had been attempted since the first edition of 1795, and therefore regimental reinterpretations abounded. An attempt by Torrens in 1824 to assimilate the drill of cavalry even more closely to that of the infantry was fortunately checked.[71] However, in 1829 Sir Hussey Vivian, then Inspector General of Cavalry, embarked on a review of the cavalry movements.[72] Included on his board were Joseph Thackwell, who had experimented in the 15th Hussars in the 1820s, and Charles Dalbiac, who, despite having charged with the 4th Light Dragoons at Talavera and Salamanca, saw himself as the custodian of Dundas's principles.[73] The results, promulgated in 1833, were certainly an improvement: the manoeuvres were simpler, officers were put at the heads of their troops, with the N.C.O.s on the flanks to preserve compactness and dressing, and when deploying on the move it was the duty of the rearmost troops to increase the pace rather than for the leaders to restrain their mounts. Nonetheless inversion remained a bogey; movements by threes persisted; and, especially when suffering casualties in action, cavalry had constantly to renumber themselves by 'telling-off' so as to regain their order. The result was that reforming took inordinately long and, as happened at Chillianwallah, the command 'Threes about!' spelt not orderly withdrawal but panic.[74] A new edition appeared in 1844, but it required the Chillianwallah débâcle, a far greater stain on the British cavalry than the charge of the Light Brigade, to ensure that in Brotherton's 1851 revision inversion was permitted.

The greater flexibility which the new regulations imparted were reflected in training that became less rigidly bound to achieving perfectly serried ranks on the undemanding land of Hounslow Heath. In Canada, Colonel Cathcart deliberately drilled the King's Dragoon Guards on broken ground,[75] and Brudenell used to take the 15th Hussars through manoeuvres at the gallop.[76] Wellington did not approve: 'All our movements are too quick for those of large bodies of cavalry; and the consequence . . . is, that in all great movements of our cavalry, they get into confusion . . . Our horse, although I believe it the best in the world, becomes unmanageable in proportion as his rate of going is increased.'[77] At Chobham camp, where admittedly the ground was very heavy after rain, the cavalry never once rose beyond a trot.[78]

As Nolan acidly observed, what right had 'they' to 'cry about English cavalry getting out of hand', if the first time they were allowed to ride at speed was in a charge in battle? It required a highly competent rider to direct a horse in full career against a compact body of men. Speed was the essence of cavalry, whether fighting or advancing into action, and cavalry training should therefore include galloping over rough country, jumping and taking obstacles – shades once again of hunting. The Duke of Cambridge, as Inspector General of Cavalry, thought likewise and instructed that more training and exercise should be done away from the school, in the open air at speed, and should include jumping hurdles and leaping ditches.[79] Nolan happily commented: 'Our riding establishments ... have much improved of late. They make the men perform the sword and lance exercise at a gallop, and teach them to take a flight of fences put up in the barrack-field.'[80]

Neither the Duke of Cambridge nor Nolan disagreed with Wellington's central point, that control and order were essential to cavalry. A charge would only succeed if the horses acted as a body. Nolan recommended that in a charge the trot should be maintained to within 150 yards of the enemy, that is within musket-shot, whereupon the gallop and the charge should be sounded in rapid succession. 'The burst, the charge itself, must always be reserved till within 50 yards', so that the horses should be well in hand at the moment of impact.[81] This was hardly revolutionary thinking: the Cromwellian cavalry, admittedly encumbered with pot and cuirass, came on at 'a good round trot', and the 1851 regulations forbade galloping till within 250 yards of the enemy, followed by 'the charge' being sounded at forty yards. All the accounts of the charge of the Light Brigade confirm the attention paid to checking the horses' pace and closing in on the centre to fill the gaps left by the dead and wounded: there is some doubt whether the 'charge' itself was ever actually sounded. The 9th Lancers at Gujerat (1849) walked to within 150 yards of the Sikhs,[82] and parts of the Heavy Brigade, whose charge was admittedly encumbered at the outset by camp lines and was over a very short distance and uphill, had not broken out of a trot when they hit the Russians.[83]

Much of the Heavy Brigade's success was attributed to the fact that its attack came from the right flank and hence took the Russians on their bridle-arms and restricted their use of their weapons. The British practice was almost invariably to place the cavalry on the flanks, a tendency justified by the national reluctance to employ massed cavalry. If the cavalry were placed in the centre, infantry had to be withdrawn, especially if they were in line, and the defensive capacity of that sector of the field thereby weakened. Infantry in column could of course operate more satisfactorily with cavalry.

Cavalry attacking from the flank were urged to do so in oblique lines or in echelon. As an aggressive move, the refused flank meant that the cavalry could hide their intended point of attack from the enemy as their nearest flank aimed for the centre. The enemy reserves would be drawn with it,

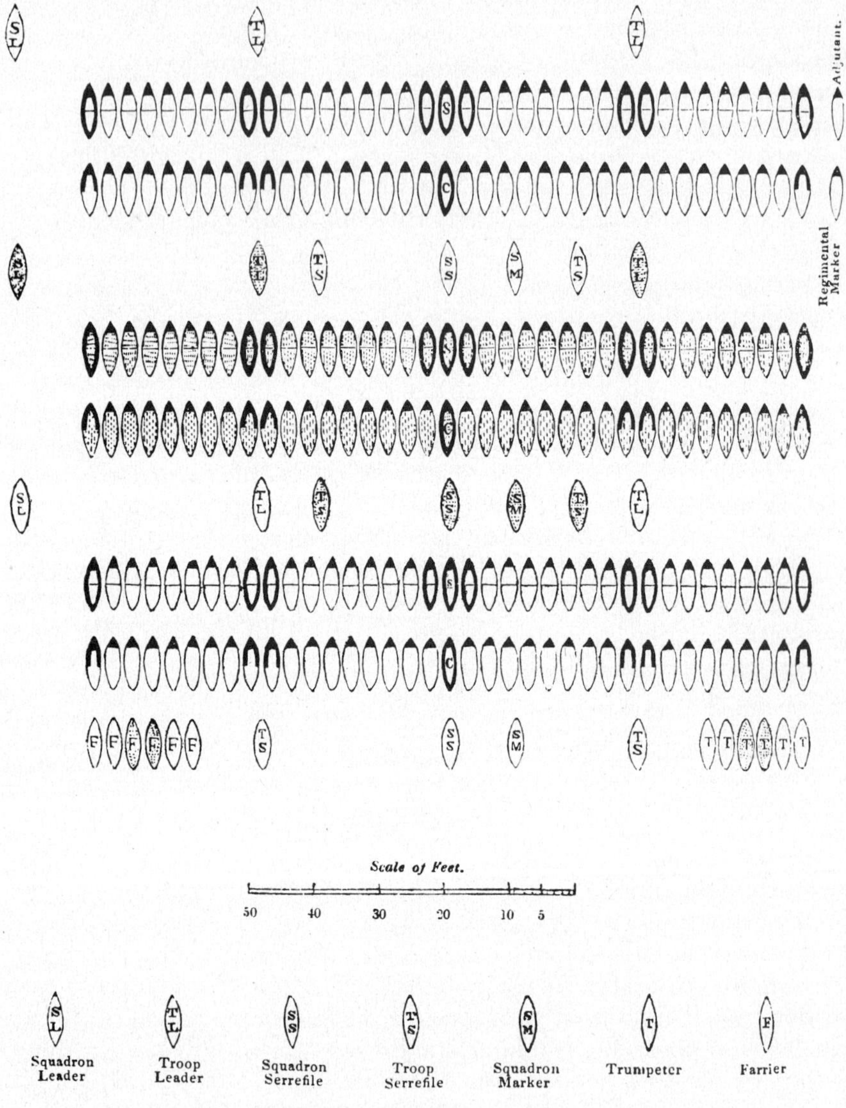
Adjutant.
Regimental Marker
Scale of Feet.
50
40
30
20
10
5
Squadron Leader
Troop Leader
Squadron Serrefile
Troop Serrefile
Squadron Marker
Trumpeter
Farrier

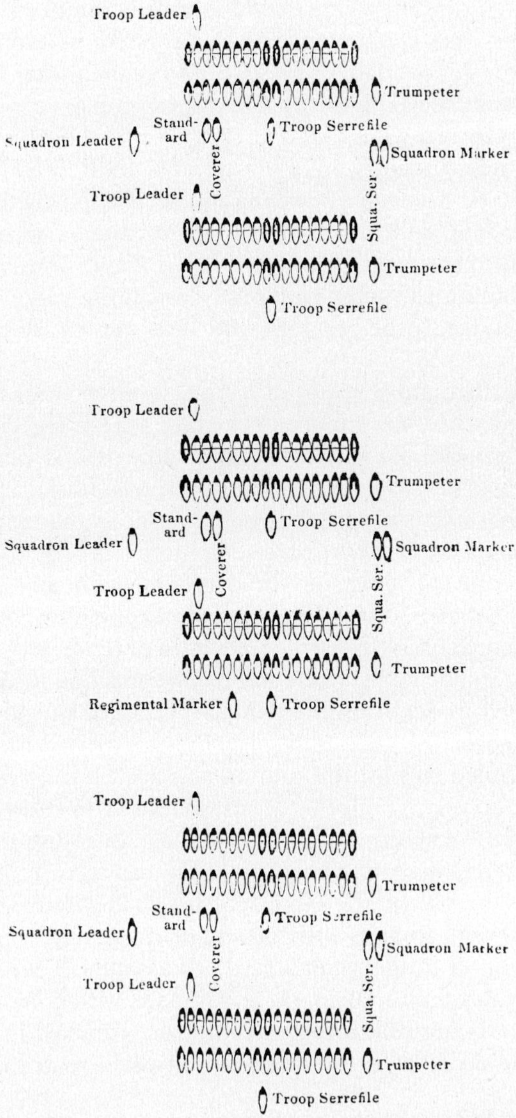


Fig. 12 a and b *The Regiment in Close Column of Squadrons* and *The Regiment in Open Column of Troops*. From *Regulations for the Instruction, Formations and Movements of the Cavalry* (1844)

thus weakening the point where the refused flank might strike. Artillery in particular were thus endangered, as the gunners would either become aware of the danger too late or be forced to realign their batteries on an oblique angle at a crucial phase. In defence, echelon gave solidarity against any attempt to turn the flank and at the same time meant that each unit could engage independently but still sure of support. Von Muller, who in an echelon attack believed in putting the weight on the outermost and refused flank, adapted the process in defence and had a massive concentration in column to the rear and outside of the flank. Thus, at the moment when an enemy attack might apparently be on the verge of success, the hidden reserves would be revealed and deal out a well-timed counter-attack.[84]

The echelon formation required a high level of training for it to be maintained under fire, and some of its virtues were in any case to be found in the simpler column. As with the infantry, the official view was that the close column was for assembly and the open column for manoeuvre, but neither was for attack. Column enabled line to be formed to the front in the quickest manner, concealed the precise strength from the knowledge of the enemy and permitted extension or deployment in any direction. The second line or 'support', therefore, might be in column so as to enable the first line, if defeated, to fall back through it. In practice, if the first line were pursued, there might not be time for the second line to deploy, but, by positioning itself in depth on the flanks, it retained the advantage of the echelon in defence.

A more extended role for the column was argued from this application to the counter-attack. Cavalry in depth was easier to control than in line, and orders more readily communicated to them. In addition they showed a smaller front to musketry fire. Sir Samford Whittingham contended that it was successive attacks on the same spot which ensured victory, and that the emphasis should consequently be on reforming and retiring as rapidly as possible in order to maintain a series of assaults.[85] Whittingham was joined by Sir Harry Smith, Bismark, and the *Aide-Mémoire to the Military Sciences*,[86] in arguing that infantry should be attacked in column, with each successive division, troop or squadron peeling off to the rear for a renewed assault.

The 1844 regulations, however, only really stipulated one formation for the attack, and that was for the first line to be in line. This in itself implied a limited role for cavalry, since the bulk of received opinion favoured the line as the best formation with which to attack only mounted troops. The normal procedure was for the front line to form in two ranks. Lord William Russell, who commanded the 8th Hussars, proposed the so-called 'rank entire' system.[87] Russell's idea won distinguished support, including that of Wellington and Vivian, both of whom agreed that the second rank following closely only served to add to the disorder – 'The fact is', Vivian pronounced, 'that the second rank is but of little use, but to fall over the first.'

General Anthony Bacon, who had been thwarted in his ambition to command the 17th Lancers by the future Earl of Lucan, tried out Russell's idea when commanding the cavalry of Don Pedro in Portugal in 1833. His success convinced Anglesey and Brotherton, and George Cathcart taught the King's Dragoon Guards to form in rank entire in Canada in 1838.[88] Bacon's disciple, Captain Kinloch, broke 300 Carlist cavalry at the head of only sixty lancers in 'rank entire'. Kinloch claimed that it doubled the effective strength of the troops involved. Herein lay its relevance for the British army. It permitted a diminutive force to present a bold front against a more numerous body with some hope of success.[89]

Russell, Kinloch and Bacon, however, were guilty of endowing the sword with more power than the horse, the individual with greater weight than the compact unit. The 'rank entire' would disintegrate passing over broken ground, and the added length of front was unlikely to contribute anything except a weak centre.[90] Certainly the 'rank entire' had a place, particularly against weak troops, and in effect the Heavy Brigade was at certain points in 'rank entire' at the moment of impact, but while cavalry still had a set role on the battlefield the rank entire smacked of unnecessary improvisation. If cavalry were on the field in strength, and it was proposed to use them in a decisive role, then there seemed little point in frittering them away in half-attacks.

For, although Bacon *et al.* claimed that the 'rank entire' made for a larger front rank, they still simultaneously endeavoured to keep a second rank in support a hundred yards away. Their confused thinking, resulting in the theoretical use of the same men in two different positions, was but a tacit acknowledgement that in a cavalry attack the most crucial force was as likely to be the reserve as the front line. In equally-matched cavalry combat, the timely introduction of the reserve might well be decisive; the staggered effect of the Heavy Brigade illustrates this and it is arguable that if the Light Brigade had had the support of the Heavies it would at least have returned with, or spiked, some of the Russian guns. A second attack on infantry might find them unloaded and still reeling from the shock of the first charge.

However, as both the Light and Heavy Brigades showed by their failure to support each other, the British weakness was to lay too much store on husbanding the reserve. Wellington favoured the cavalry's being in three stages, each 400 yards apart; Hussey Vivian countenanced this arrangement even for a force as small as three squadrons,[91] and official regulations were similarly over-cautious. The 1851 edition stipulated that the 'support' (the second line) and the 'reserve' (the third) were both to be in column, not line, and that the line itself should never be more than a third of the total force. Certainly these provisions obviated the danger of losing the whole cavalry force in one rash enterprise, but they also left the striking power so weak as greatly to diminish its capacity for decisive action. The rare occasions when cavalry did operate to effect on the battlefield testify to how often they sat frustrated, their divisional commander, probably an infantryman, distrusting their over-eagerness and imagining unknown dangers.

And yet the cavalry, although gaining in professional self-regard in the years 1815–54, had not really come up with a solution to infantry firepower. While the smoothbore musket remained the dominant weapon, Nolan and his ilk were of course right – good leadership and determination, faith in the ability of cavalry and its employment at the crucial moment, meant that its role in the set battle could be preserved. But the adoption of the rifle left it sinking where before it had only been floundering.

Le Couteur calculated that with the Enfield each man could fire twelve rounds at an approaching cavalry charge, and given the large size of target this figure was tantamount to twenty.[92] Furthermore each round was better aimed and more penetrating than a Brown Bess ball. Nolan had no real answer to this: he confessed that the new inventions were 'not to be despised', but the nearest he came to a recognition of the truth was that 'The great improvement made in firearms and the increased range of the infantry musket, leave but little chance for cavalry, unless the speed with which they can pounce upon the infantry lessens the number and the effect of the discharge to be received during their advance.'[93] Galloping from 800 yards would destroy the compactness that was meant to be the key to a successful attack against infantry: assuming that the cavalry took two and a half minutes to cover the distance, each trooper would probably have ten rounds directed at him. The impetus of the charge would be lost long beforehand. Although many practical lessons would have to be dealt out first, to the percipient the message was already clear – heavy cavalry, in their traditional role as shock troops, were finished.

However, cavalry in the broader sense were not. For those who wished to see, the need for the horse on the battlefield was bound to be continuous until a superior form of rapid transport emerged. Even on the set-piece battlefield, a flanking movement or intelligent use of cover might bring dividends for the resourceful commander. In open country the roles of mounted infantry, skirmishers, harassers of convoys, scouts and advance guards remained inviolate. All cavalry now had to undertake the duties previously considered as confined to light cavalry. Owing to its colonial commitments and to its small size, the British cavalry was already far advanced in this direction.

### (ii) Other roles

One single contribution above all others – perhaps in itself a reflection of the doubts of the Peninsular school – dictated the pattern of cavalry tactics and indeed the development of the arm itself. By the yardstick of European armies, the numbers of the British cavalry were contemptible. In 1852 J.-W. Jervis showed that the British cavalry was equal in strength to an eighth of its infantry force: in Prussia and Bavaria this figure was a quarter, in France and Austria a fifth and in Russia a sixth.[94] Therefore, according to the conventions of European warfare, the British mounted arm was woefully inadequate in proportion to its infantry. Moreover, since the whole British army was between a third and a quarter the strength of the French, the absolute numbers of its cavalry were diminutive. This numerical deficiency had been a prime cause for the cavalry's limited achievements in the Peninsula: in 1827 even Wellington called for an increase in its strength.[95]

Tacit acknowledgement that the peacetime strength of the cavalry was too low was shown in the high number of regiments (twenty-seven) relative to the arm's overall establishment. It was argued that this arrangement provided cadres ready for rapid expansion in war.[96] The line cavalry regiments, with the exception of the King's Dragoon Guards and of the regiments in India, each mustered 304 rank and file but had a disproportionate number of officers (twenty-seven). The regiments in India, which were effectively on a war footing, had an establishment of 675 other ranks and forty-seven officers.[97] This virtual doubling in size put unnatural strains on a unit. On a personal level twenty officers per regiment were likely to have to go on half-pay on the regiment's return from India, and many more seasoned soldiers to be discharged. The complexion of field manoeuvres was totally transformed by the difference in the strength of troops. Furthermore, in the event of an emergency, expansion could hardly be rapid when only 271 of the 304 were mounted, and much time and many men would have to be devoted to the training of remounts for the augmentation. Bismark spoke of between 750 and 1,000 sabres per regiment as an ideal,[98] and even British officers suggested figures between 560 and 800.

Indian service, while not obscuring the fragmented nature of the cavalry's organisation, did hide the overall low numbers of the arm. Thanks to the East India Company's contribution four large brigades of cavalry were

Fig. 13 *Royal Horse Guards-Blue*: the ideal of heavy cavalry, complete with cuirasses, charging at Chobham camp in 1853. Aquatint by J. Harris after H. Martens (National Army Museum)

attached to the army of the Sutlej in 1846, and regular reliefs meant that in 1849 in the 2nd Sikh War the 13th Light Dragoons still mustered about 600 men.[99] The hurried equipping of an expeditionary force from the United Kingdom revealed a totally different picture. Regiments had to go to the Crimea at strengths lower than even the English establishment, since two depot troops were left behind to train remounts. The overall strength of the five regiments of the Light Brigade was fixed at 1,250, with 1,000 horses,[100] the 11th Hussars, for example, having two squadrons of 125 each. Even as early in the campaign as Balaclava, the rigours of service allowed the 8th Hussars to put only 104 into the field, and the 17th Lancers, the strongest of the light regiments, had a mere 145 men.[101] Lucan wanted squadrons of 170 men each, with three to four squadrons per regiment, giving him a realistic, but hardly ideal, force of 3,600 sabres.[102] The bulk of informed military comment envisaged an important role for a large cavalry force in the campaign, and the Duke of Cambridge, mindful perhaps of the potential of cavalry in Turkey (the original seat of war), cried out for at least 10,000.[103] In fact, Raglan sailed with 2,000[104] and at the Alma had only the Light Brigade with him: any criticism of his failure to prosecute his victory must be tempered by the fact that a mistake at that juncture would have left him in unfamiliar enemy territory without the 'eyes and ears' of his army.

Such a pitiful supply of horsemen clearly inhibited any ambitious employment of cavalry in the field. It also discouraged over-specialisation when slender resources were likely to dictate that all cavalry be capable of all duties. Nolan, reflecting the European ideal, saw the cavalry as falling into three distinct groups, 'not one of them being fit to perform all the duties required of horse soldiers in the field'.[105] The first category, true heavy cavalry, of which Napoleon's cuirassiers were the paragons, found no place in the British army. The three regiments of the Household Cavalry alone wore body armour and it seems clear that even they would have discarded it if they had gone to the Crimea.[106] In any case the cuirass impeded the soldier's movements in a mêlée and added extra weight to the horse, whose essential strength, in the face of the new infantry fire, had to be speed. The role of the heavy cavalry was shock-action against infantry squares; the Minié ensured its death. Next to heavy cavalry Nolan placed dragoons, traditionally mounted infantry, but by the nineteenth century a conventional description for cavalry geared to most of the arm's duties and in particular an apt epithet for the bulk of the British cavalry of the line. The dragoon, in British terminology, was a heavy cavalryman. Finally he listed light cavalry, 'the most important in the field'. Their main duties were skirmishing and reconnaissance, but in addition their manoeuvrability enabled them to tackle their heavier counterpart.

The European tradition, as Nolan's categorisation testifies, clung rigidly to the heavy and light cavalry distinction. 'No less serious', Jervis wrote in his edition of Lallemand, 'are the inconveniences resulting from one corps doing the duty of another; as, for example, hussars doing the duties of cuirassiers, or dragoons that of lancers: for the spirit of those different branches of the services neither can nor ought to be the same'.[107] Such scruples could hardly be allowed to impede the duties of a colonial commander in the field. At the Cape, the 12th Lancers had to forsake their lances for carbines. Even in India, where cavalry were more plentiful, the 9th Lancers requested to be allowed carbines to enable them to meet Sikh matchlockmen, thorough-going light cavalry not apparently having been available.[108] Bismark proposed a deceptively appealing solution, namely that all cavalry regiments should be of one type, four squadrons being formed of line cavalry and four divisions of light skirmishers.[109] The compromise, theoretically ideal for isolated regiments overseas, in fact depended on a far larger regiment than the impoverished British establishment would allow.

Specialisation was further discouraged in the British service by the conclusion that heavy cavalry were unfit for colonial service. Nolan, stating a truth too often negelcted in the British army, saw that it was not his uniform or equipment that really made a cavalryman heavy, but the power and physique of his horse.[110] Hardy, nimble mounts were the requirement for the fatigues of outpost duties and warm climates. Even in the temperate zones of Canada, Charles Grey had had to modify his praise of Cathcart's King's Dragoon Guards by noting that their 'horses were too fat to work'.[111] The Canadian rebellion in 1838 was an exceptional instance: it was the only occasion until 1854 that horses were transported by sea. Normally regiments ordered to the colonies took over the mounts of those they were relieving. Therefore horseflesh ought to have been no impediment to heavy cavalry sailing abroad. However, the native Indian remount was a lightly built, if hardy, beast, perhaps boasting Arab blood. The large man beloved of commanding officers could not be mounted in India; thus the 6th Dragoon Guards postponed their turn of service in 1845.[112] The 7th, who were sent to South Africa in 1843, had similar difficulties with the Cape horse: the lightest man weighed 10 st 6 lb and was 5 ft 7½ in tall,[113] with the result that 'No well fed horse can carry them: much less a horse frequently working on grass. The Land's End is quite far enough for a five foot ten dragoon to move to.'[114] Consequently, except for the 7th and the King's Dragoon Guards, no heavy cavalry (three regiments of Household Cavalry, seven of Dragoon Guards and three of Dragoons) went abroad until the Crimea, and none to India until 1857.

Therefore quite justifiably, and certainly not uniquely, the Deputy Adjutant General castigated the heavy cavalry as 'nearly useless to the country'.[115] The burden of foreign service fell squarely on the thirteen regiments of light cavalry (all officially light dragoons, but five equipped as hussars and four as lancers). Idle though they were, the heavy cavalry regiments were protected by the arguments of those trained in European war, and those fearful of a French invasion. Nonetheless even the most ardent Europeans had to agree that the heavy cavalry were proportionately too strong. Jervis thought they should muster a quarter of the strength of the light cavalry:[116] in fact the two were roughly equal in size. The Duke of Cambridge, as Inspector General of Cavalry in 1853, argued that 'the first Military authorities on this subject' favoured the retention of both types, as 'for home service whether for purposes of defence or in the event of a Continental War, a due admixture of Heavy and Light Cavalry would be most desirable'. But he acknowledged that for foreign service 'the only description of Cavalry that can really be made available is Light Cavalry', and therefore concluded that to obtain a reasonable rotation of service the heavy cavalry of the line should be reduced to six regiments, the remainder being made light.[117]

Colonel Luard contended that in such a small army the distinction of light and heavy should be abolished altogether.

The horses of our cavalry differ a little in strength, but all are equally good, and of one description. The men are recruited from the same class of people, and though varying in weight, are all sufficiently strong, and equally brave and active; both heavy and light are instructed in outpost duties and in skirmishing, and are expected in the charge to overthrow any cavalry that can be brought against them ... There is no difference in the duties of our cavalry ...

only their dress distinguishes them.[118] Colonel Leach and other writers in the military press agreed.

In practice, the distinction between heavy and light regiments did become blurred. Since Dundas's time, the cavalry manoeuvres had made no distinction as to type, and it was enunciated that the principles of military equitation 'are the same for all classes of Cavalry'.[119] The horses of regiments sent abroad were distributed equally between heavy and light regiments.[120] The 1844 *Queen's Regulations* expressly stipulated that, as the cavalry force was proportionately small, 'this portion of the army should be of the *best* description, that is, that both the *Heavy* and *Light* Cavalry shoud be equal to the *Charge in Line*, as well as to the *Duties on Out-Posts*'.[121]

The official tendency, therefore, was to make all cavalry of one class. However, that class was heavier than the ideal light cavalry. *Queen's Regulations* went on to conclude that 'The Horses, which are selected and trained for the Cavalry, should therefore, be of a sufficient height and strength'. The excellence in the Peninsula of the King's German Legion's light cavalry led Wellington to the conclusion, hinted at in the 1844 *Regulations*, that 'we ought to trust to the foreign cavalry with the army for the ordinary outpost duty, when the employment of our own excellent cavalry on such service should not be necessary'.[122] British horses were strong, and, 'deriving their power from cart-horse blood, have not the breeding and activity required for light cavalry duty'.[123] The men were brave. The cavalry were therefore better suited for heavy roles, and in particular for the charge. As late as 1855 Hardinge argued that the size of both horse and man made heavy cavalry Britain's natural forte.[124]

Certainly many British light cavalry regiments rode very heavy. In marching order, the weight of a heavy cavalryman (put at 11 st 8 lb) and his equipment totalled 18 st 10 lb, while his lighter counterpart (estimated at 10 st 3 lb) actually totalled 18 st 4 lb when astride his steed.[125] The 9th Lancers, on being sent to India in 1842, rode nearly 19 st, too heavy for the native horses to bear them.[126] In 1846 the 10th Hussars, shortly before their departure to India, were inspected by General Brown, who found them for height and size much more like 'Lifeguardsmen than Light Dragoons'.[127] The 12th Lancers at the battle of Berea, at the Cape in 1852, were found 'too big for their horses & too slow'.[128]

It was the native British horse which encouraged the light cavalry to ride heavy and gave them their ability to charge in line. But such an animal was hard to supply and feed on campaign. Trials at Bombay had favoured the virtues of the more adaptable Cape remount,[129] and Nolan called for the introduction of Arab and Persian blood to give the English stock a compact form and wiry limb.[130] However, a lighter horse meant a lighter rider, and although a maximum of 5 ft 8 in was introduced for the light cavalry, it was not uniformly observed.

The potential virtues and vices of this hybrid light cavalry were revealed in the Crimea. India, where reconnaissance and skirmishing were much better done by native regiments, had encouraged the light regiments to favour a heavy tactical role – as the 16th Lancers had done at Aliwal. The charge of the Light Brigade, tactically a heavy cavalry action, was therefore a logical growth from Indian experience as well as the fruit of revised views at home. However, outwith the set battle, the horses were unfit for the rigours of a long campaign. Only the Arabs proved resilient in the winter of 1854–5. On his return from the East, Sir John Burgoyne publicly bemoaned the British lack of a genuine light cavalry, well adapted to outpost work.[131] The conclusions of George B. McClellan are particularly interesting, if only because the United States had but one form of cavalry, and they were operating in the not dissimilar conditions of the western frontier. 'The horses of the English cavalry', he wrote,

> are large and excellent; for the heavy cavalry they leave nothing to be desired; it may be a question whether they have light cavalry, in the true sense of the term, except perhaps, some of the regiments who have been serving in India, and are mounted on Indian horses; for the men and horses of the light cavalry are scarcely to be distinguished from those of the heavy, and it may be doubted whether they would stand the severe work, exposure, and short rations, which usually fall to the lot of light cavalry in campaign.[132]

However, the process also operated in reverse. Arguably the heavier cavalry was becoming lighter. The transformation was initiated by the constant outpost work in the Peninsula, and was confirmed by the exigencies of colonial garrisoning. Although no heavy regiments were made light in the way some would-be reformers advocated, in 1842 Wellington decided that, out of fairness to the light cavalry, a heavy cavalry regiment, the 7th Dragoon Guards, should be sent to the Cape.[133] In 1851, the 6th Dragoon Guards, in preparation for foreign service that was eventually postponed, were re-equipped as light dragoons, producing a hybrid uniform that is still cherished.[134] As a result by 1853 the Duke of Cambridge could speak of both the 6th and 7th Dragoon Guards, and also of the 6th Dragoons, as effectively light regiments.[135] Hardinge endeavoured to recruit smaller men for the heavy cavalry, eventually setting a maximum standard of 5 ft 9 in.[136] Certainly, in the Crimea, the Light Brigade's exploits in the charge can be contrasted with such examples as the 2nd Dragoons skirmishing on foot at MacKenzie's farm or the 4th Dragoon Guards acting as vedettes on 28 October 1854.

Therefore, the heavy cavalry had by dint of colonial service come some of the way towards accepting the role which infantry firepower after 1850 was going to compel all mounted troops to adopt. Built on the traditional foundations of light cavalry work – skirmishing, outposts, advance and rear guards, and reconnaissance, it ensured a future for cavalry so long as the horses provided greater mobility than any other form of locomotion.

'The chief duties of light cavalry', according to one of Lucan's divisional orders in the Crimea,

> are by their vigilance and activity to ensure the safety of the army from all surprise, and to procure all the information within their power respecting the enemy and the country; but it is not their duty needlessly, without authority, to engage the enemy. When attacked, it is doubtless their positive duty to defend themselves at all hazards; but on no account should any party attack or pursue, unless specially directed so to do.[137]

Reconnaissance or outpost duty therefore carried an implicit denial of combat. It was unlikely to appeal to the ambitious young cavalry officer, and it conflicted with the emphasis on resolution in the charge. Moreover, the troop or squadron commanders, who would be acting in an independent role on service, were vouchsafed few opportunities to practise their skills. In Britain the training grounds were not available, and in India native regiments were likely to prove more adept in gleaning intelligence. At field days, drill and reviews were the favourite occupations, and command was rarely devolved from the commanding officer. In 1853 Fitzclarence instructed the 10th Hussars to reconnoitre and produce military reports on the land round Kirkee,[138] but this was a rare example.

Furthermore, practice was not encouraged by the prevalent view that light cavalry duties could only be learnt in the field, and that in the absence of active service recourse should be had to manuals. Prompted by the fact that the cavalry regulations were as thin on the duties of light troops as were their infantry counterparts, the unofficial works on outposts and reconnaissance were thorough and precise. A number of cavalrymen committed themselves to paper, many of them drawing on the work of von Arentschildt, whose book was written in 1809 and was still recommended, in the abridgement of Ponsonby, to the cavalry in the Crimea by Lucan.

Fig. 14 *13th Light Dragoons in Pursuit of the Enemy*: light cavalry patrolling, with carbines ready. Aquatint by J. Harris after H. Martens, published 1853 (National Army Museum)

Arentschildt was an officer of the King's German Legion, and the defensive nature of the early years of the Peninsular War is abundantly clear in the limited scope of his instructions. The result was a very static interpretation of light cavalry duties. In advance, rear or side guards, a large reserve should be nearest the main body, with the support, advanced parties and foremost files radiating from it. Similarly, the inlying picquet was succeeded by an outlying picquet, vedettes and, further on, patrols. The proportions of each group were calculated in accordance with the size of the force they were guarding or operating with, and the distances between the layers were finely calculated so that each could remain in contact and at the same time give sufficient warning of an enemy approach. The commander of the outlying picquet was to devote his time to planning his defence and noting adjacent points of tactical value. All these instructions were of prime importance to the novice, but made remarkably little concession to the fact that a cavalryman had a horse. One of von Arentschildt's followers, Lieutenant Robbins, suggested that a picquet should be posted on the rear of a hill, wood or copse, near a bridge or defile: for infantry this was orthodox thinking – it ensured a safe retreat, hid the picquet from the enemy, and provided a place to make a stand should it be required – but it was strangely inflexible to tie the mounted arm to such obvious geographical features.[139]

The result was an unenviable reputation for British light cavalry in this, the foremost of their responsibilities. The amazing story of the swing around Sebastopol to its southern front, when both armies passed so close that it was left to the British Commander-in-Chief and his staff to discover the Russian tail, points primarily to bad reconnaissance and, despite the small numbers of allied cavalry, remains inexcusable. Even in India, where the virtues of native irregular cavalry could have been exploited, inadequate scouting had let the British be surprised at Mudki.[140] Sergeant Pearman's account of the picquets at Ramnagar in 1848 suggests that it was not neglect of the manuals that led to these deficiencies: a double chain of infantry vedettes lay within a ring of cavalrymen, two stationed at every 100 yards, and the whole army was turned out an hour before dawn.[141] However, the intelligence gathered by a thin and static chain of guards was limited to the vision of the outermost man, which in bad weather might be only a few yards. The answer lay in patrols, not confined to liaising between existing picquets, but employed on long range reconnaissance and probing deep into the enemy's territory.

Even von Arentschildt had advocated 'patrols of discovery' and 'secret patrols', which were small bodies to travel far into the enemy rear for great lengths of time.[142] Captain Short in his *Treatise on Patroling* (1831) listed small, ordinary and 'Grand Reconnoitring Patrols'. The latter could be up to squadron or company strength and would be employed at corps or divisional level to establish broad questions of enemy strategy, to protect the flanks, and even to threaten attack at unexpected points. Although he spoke of a corps of observation, with strong patrols detached in a radius of over twenty miles around it, he did not emphasise its potential in war, and both Short and von Arentschildt were mainly talking of small contingents with limited objectives. However, on similar duties Napoleon had employed large bodies of cavalry, and the Light Brigade's reconnaissance to the Danube in 1854 also smacked of a long-range thrust. The uhlans in the Franco-Prussian war proved that bold reconnaissance still provided cavalry with an important role in the age of the breech-loader.

Operating in isolation compelled light cavalry to be self-sufficient in armament, and to be able, if forced to fight, to skirmish. For this aspect of their duties, the British cavalry were better attuned. The 1833 cavalry regulations made provision for the first time for whole squadrons to skirmish. They recognised that, perhaps even more than light infantry, light cavalry skirmishers enabled a commander 'to conceal the strength of his troops and to reconnoitre that of the enemy, as well as its position and the ground immediately in front of the line of skirmishers'.[143] For all these duties – the skirmish, the picquet and the patrol – the appropriate weapon was not the sword, but the carbine.

Although not yet formulated as a direct challenge to the *arme blanche*, the equivocation of the lancers between the carbine and the designated weapon of their arm is evidence of the former's potential application to all cavalry. It was extremely difficult to carry a lance slung on the arm and a carbine fastened to the pouch-belt, and it defied human ingenuity to handle both when mounted. Trials initiated by an optimistic William IV in 1834 confirmed this.[144] Nonetheless, carbines were approved for the use of lancers in 1837.[145] The intention was presumably to issue them in case of need rather than to expect their regular employment. However, one account of Aliwal mentions the 16th Lancers firing carbines,[146] and double-barrelled rifle carbines were issued to the 12th Lancers at the Cape.[147] The instances noted of lancers using carbines were presumably exceptional, but they do point not only to doubts about the true role of lancers but also the favour with which the carbine was viewed.

This is all the more surprising in that the British carbine remained a wretchedly constructed weapon. Improvements, principally in the lock, were effected by Moore and Manton between 1829 and 1831, and in 1833 Hill approved Manton's version for heavy and light cavalry.[148] From 1832 experiments were conducted on a percussion lock for cavalry, and the reports from the King's Dragoon Guards and 7th Hussars in Canada were so favourable that in 1839 the so-called 'Victoria' carbine was adopted.[149] In 1843, its back-action lock was replaced by a side-action one.[150] The improvements in the weapon were more than counterbalanced, from the cavalry point of view, by the decision that all British military firearms should be of musket bore. This meant that the shorter barrel of a carbine caused an intolerable recoil unless the charge was reduced. But if the charge was reduced, so was the range. Nonetheless the barrel still weighed as much as 7 lb 9 oz and was three feet, six inches long. The Horse Guards acknowledged that the new carbines could not be fired in one hand like a pistol but claimed that 'if the Dragoon is trained to rest his carbine on the wrist of his bridle arm, the new carbine becomes a formidable and highly efficient weapon'.[151]

In 1855, the Board of Cavalry Officers more honestly concluded that the Victoria carbine 'is too heavy and clumsy to be used on horseback; fit only for dismounted duty, for which a lighter and smaller weapon of proper construction, would serve equally well'.[152] To achieve accuracy of shooting and lightness in construction, a smallbore rifle was clearly desirable. However, muzzle-loading was difficult enough on horseback without adding to its problems by rifling the barrel, and clearly the desideratum was a breech-loading mechanism which would be more easily operated on horseback.

As early as 1841–2, a breech-loading carbine of American invention went as far as a regimental trial.[153] Some regiments had the rifled version of the pre-1833 Paget carbine, which with a bore of .685 in and a weight of 5 lb 6 oz was at least handier for a horseman to use.[154] Nonetheless the Paget was unsuitable for employment in the Crimea since it demanded a different type of ammunition. Hardinge was so opposed to the heavy Victoria carbine that in 1853 its issue was suspended pending further trials and inventions,[155] and in 1855 he directed that carbines be bought direct from America rather than revert to it.[156] However, the technological development of rifled breech-loading carbines proceeded so slowly that supplies of any weapon in the years 1854–5 were scarce.[157] Trials were conducted in 1854 on the arms submitted by Sharp and Leitch, and in 1855 the Board of Cavalry Officers reported in favour of the latter, which was so light as to justify the description 'pistol-carbine'.[158] Hardinge, although aware that a better weapon might soon materialise, endorsed the Board's views and 15,000 were ordered.[159]

Leitch's carbine was only accurate to 300 yards. One of the agonies surrounding carbine development arose from the fact that cavalry were not exactly sure what was to be the application of the weapon they were seeking. Was it to be accurate at long ranges, able to compete with infantry fire and probably therefore confined to dismounted action, or was it to be light and handy, fired and cocked with the motion of one hand, the bridle-arm being sufficient support for it to be fired mounted? The solution to the quandary lay in the decisions as to the future role of cavalry on the battlefield. While the horse remained the cavalryman's primary weapon, the carbine was to be an ancillary arm. If the value of the horse was as a means of rapid transport, then the carbine could afford to be heavier, designed to take for itself a more dominant role in war.

Consequently, for the Peninsular school, the carbine was a weapon for mounted vedettes and was certainly not expected to exert any decisive influence in battle. Even in 1855 Hardinge could comment that 'the carbine is not the arm on which cavalry depends for its defence – it being applicable only to vedettes and outpost duty, and being intended there for the purpose of giving timely notice to the picquets and troops in the rear, than for permanent resistance'.[160] The regulations stated that 'the fire of the Cavalry Soldier is never to be had recourse to but in skirmishing'.[161]

This limited appreciation of the potential of the cavalry carbine stemmed from a widely held belief that it could not be fired efficiently while astride a horse. A cavalry skirmisher could not perform his duties adequately unless he was mounted, but this demanded that at the moment

Fig. 15 *Fighting in the Bush*: the Cape Mounted Rifles firing from the saddle against Kaffirs, as depicted by Thomas J. Lucas, a captain in the Cape Mounted Rifles, in *A Campaign in South Africa* (1861) (British Library)

of firing the horse should be stationary. The 1833 *Cavalry Regulations* could offer no solution to the danger, apart from stating it:

> Although it is desirable that the horseman should be habituated to the use of his carbine at speed, few occasions can arise for his using it against an enemy at any pace beyond a walk; and notwithstanding he is enjoined, while skirmishing, to keep his horse in motion, in order to avoid becoming a fixed object for the enemy's marksmen, he cannot reasonably calculate on his shot being effective, unless he halts for the moment of firing.[162]

Even so, the flash and noise of the discharge were liable to upset the horse, whether moving or not, and the animal tended to bring his head up or to shy at the crucial moment. It was easy to conclude that the carbines of the light horse were calculated to do little better than 'frighten the crows'.[163]

The truth of these verdicts is hard to assess. The attention to ball practice was so poor that the motion of the horse had as good a chance of aiding the firer's aim as upsetting it. A dismounted troop of sixty men, armed presumably with the overweight Victoria carbine, only put eighteen shots out of 180 on a target six feet by two feet at 100 yards.[164] In fairness, eighty yards seems to have been reckoned the effective range of the carbine, and, when the annual ammunition allowance only totalled ten rounds of ball and fifty of blank, little opportunity existed for an improvement in the mounted man's shooting. The high proportion of blank at least presented an opportunity to accustom the horse to the discharge. The Cape Mounted Rifles, admittedly with a different pattern carbine, are alleged to have fired from the saddle even in the charge, and were trained to do so.[165] The 1844 *Instructions for the Carbine Exercise* stipulated practice at a specific target on a horse in motion. That such training might well prove fruitful was demonstrated by the adjutant of the 10th Hussars who at 200 yards could hit the figure of a man with a rifled carbine when mounted.[166] The adoption of Leitch's 'pistol-carbine' in 1855 was confirmation that official policy expected the horseman to be able to fire when mounted.

Even should the horse become accustomed to his rider's firing and the soldier himself prove reasonably proficient in shooting, it was clearly necessary, if any aspirations to accuracy were to be retained, that a skirmisher should rein in before firing. The comparatively large surface area of a mounted man was bound to be a target for marksmen, and such a pause, particularly in an exposed forward position, might well prove fatal. Therefore, although proficiency with the carbine on horseback was emphasised in the early 1850s, the obvious procedure was for the cavalryman to dismount to fire. This ensured the best use of cover, the fewest impediments to accuracy, and could allow a heavier and longer-ranging weapon to be carried.

Such, in a sense, was the traditional role of the seventeenth- and eighteenth-century dragoon, who in origin was no more than an infantryman mounted on a horse for the purpose of rapid conveyance to wherever he was most needed. However, by the start of the nineteenth century, the British dragoon had become too imbued with the cavalry spirit to show much enterprise as an infantryman. The 1833 cavalry regulations instructed the cavalry soldier how to drill on foot but were thin on his duties as dismounted skirmisher. The 3rd Light Dragoons skirmished on foot with carbines at Gujerat,[167] but most commentators felt that a cavalryman without his horse was weak opposition indeed. The calls of men such as Colonel Leach and Sir J. E. Alexander for the British army to imitate the French *chasseurs à cheval* is an indication of the lack of such an arm.[168]

Both Leach and Alexander were well aware of the requirements of a reduced colonial army. By contrast, observers of the European states could see the eighteenth-century dragoon still extant, particularly in France and Russia, and they concluded that he was little better than an unsatisfactory hybrid. Beamish, in his edition of Bismark, argued that properly trained infantry must fight better than mounted infantry,[169] and Nolan criticised the mounted infantryman as being fit only for light infantry duties and making an inadequate assault trooper.[170] These condemnations of the old-style dragoon had the support of Wellington, who observed that while the British cavalry's firearms were inferior they could not be pitched against the foot soldier.[171]

Such criticisms were partly based on a European misunderstanding of the mounted infantry's role. Russia attempted to make a cavalryman behave like a 'heavy' infantryman, and in France a dragoon was expected to skirmish and flit from cover to cover while encumbered with boots and helmet. An observer of the Chasseurs d'Afrique, who so cleverly extricated the remnants of the Light Brigade by dismounting and using their long-barrelled carbines, would have seen in them the precepts learnt in colonial warfare. Sir J. E. Alexander, reporting on the United States cavalry, whose duties bore comparison with those of the British cavalry, called for 'a couple of regiments in England, to act on foot and on horses' backs, as mounted rangers'.[172]

Therefore, while the European school in Britain tended to denigrate the employment of mounted infantry, those with colonial experience were more flexible. A smaller body of mounted men could garrison and patrol vast tracts of country more efficiently than a larger force of infantrymen. But much of the ground was either broken and heavily wooded as in Canada or covered in bush as in South Africa. In this terrain skirmishers with accurate firearms were wanted. Mounted infantry therefore filled a military necessity in colonial terms. It also enabled an understrength army to employ its nearest available troops on almost any duty. A positive feature of the similarity of the cavalry and infantry evolutions was the ease with which one could adapt to the other.

Mounted infantry appeared therefore as *ad hoc* fulfilments of specific local requirements and were not regarded as a regular part of the British army. Infantry were horsed to undertake police duties in Australia,[173] and in Scinde Sir Charles Napier formed a camel corps, with a driver and infantryman on each animal, capable of covering a hundred miles in forty-eight hours. Alexander and Captain Robertson, alias 'Ubique', repeatedly called for a properly organised corps of mounted rifles for service in North America.[174] The two cavalry regiments sent to Canada in 1838, the 7th Hussars and the King's Dragoon Guards, were issued with hatchets and the new percussion carbines.[175] George Cathcart, the commanding officer of the latter regiment, stressed the skirmishing and outpost duties of even nominal heavy cavalry in the broken terrain, and concluded, in orders of 1840, that, as cavalry could move faster than infantry, they should take up a position with greater promptness, dismount, and hold the ground until the infantry came up to relieve them.[176]

When the 7th Dragoon Guards were ordered to the Cape in 1843 they exchanged their carbines for long rifles and bayonets.[177] Although they had been rapidly trained in their use by the Rifle Brigade before sailing, the rifles remained something of a mystery to the 7th. Only practical experience in the field taught the regiment that loading them on horseback was impossible, and in 1848 the Horse Guards were belatedly forced to observe that 'it never was proposed or expected, that they were to fire, much less to load these Rifles on horseback, but to ride up and to fire on foot, or across their saddles whenever they could contrive to get up within range'.[178] By 1845 the regiment had learnt to dismount to fire and to fight in the bush as infantry. So far did the 7th progress along the path to being full-blooded mounted infantry that, on their return, the Inspector General of Cavalry disregarded complaints about their equitation and instead recommended that some additional infantry officers be drafted into the regiment.[179]

In view of the dearth of cavalry, a more customary expedient at the Cape was to mount a company of foot rather than dismount a troop of cavalry. In 1835, the light company of the 75th was equipped with carbines and mounted on 'small but hardy and strong' horses for duty at the Cape. Sir George Napier reported in 1838 that it was 'the most efficient mounted force I possess'. However, at the end of the same year Howick ordered its return to normal duties, on the grounds that too much training was required to turn a foot soldier into a cavalryman and that he was then despondent when deprived of his horse.[180] By 1842, Napier was again bemoaning the lack of cavalry and requested authority to mount one company of each infantry regiment as dragoons.[181] His proposal does not seem to have come to fruition, but in 1846–7 the Rifle Brigade mounted a company for service at the Cape,[182] as did the 45th Foot in 1849–50.[183] In the Eighth Kaffir War, several bodies of mounted infantry seem to have been formed, including contributions from the 43rd[184] and 74th Foot.[185]

The model for all these arrangements was the Cape Mounted Rifles. Although a local corps in that its services were confined to South Africa, it was the only unit in the army which could officially be described as mounted infantry, that is infantry with cash allowances for horses. Formed of Hottentots, with knowledge of the area and of the Kaffirs' ways, it was equipped with a serviceable green uniform and armed with a double-barrelled carbine. Swords, if issued, were not used even in the charge, the men being trained to fire their firearms from the saddle. First made in 1822, the carbine boasted twenty-six-inch barrels and a calibre of .733 inches. It was converted to percussion in 1840[186] and in 1851 trials were held on a rifled version, sighted to 800 yards, which was issued to the

12th Lancers for the Cape in 1852[187] and to the Cape Mounted Rifles themselves in 1853.[188] With a weight of ten and a quarter pounds and a fearsome recoil, it was clearly a formidable weapon to handle on horseback. But the reports of infantry officers engaged in the Seventh Kaffir War show that the carbine was highly regarded: Lieutenant-Colonel Michel said that 'such a noble instrument I have never seen in the hands of a soldier before'.[189] These accounts make it clear that the double-barrelled carbine was peculiarly a mounted infantry weapon, its weight, whilst not precluding its use on horseback, encouraging the Cape Mounted Rifles to come to grips on foot, where its short barrels and two shots gave it an advantage in the almost sporting circumstances of unexpected individual encounters in the bush. One man in three would hold the horses while his comrades dismounted to fire or skirmish on foot. A typical account is of their conduct at the capture of Fort Armstrong in 1851: the Cape Mounted Rifles 'galloped up the face of the hill and jumped into the midst of the enemy; then dismounting from their horses they attacked the enemy, who continued firing from huts and concealed positions, and drove them from every point'.[190]

The adoption of the Minié uprooted the conventional use of the mounted arm from the battlefield. But colonial experience had already given the British army an inkling of the cavalry's future. Chesney cited the Prussian, Captain Wittich, to the effect that more cavalry should have rifles and be trained to regard the horse only as a conveyance: the 'cavalry spirit' should be allowed to die. He observed that the Cape and India had shown the value of mounted infantry, and that the time had now come for all cavalry to have breech-loading carbines and to learn to skirmish on foot or on horseback.[191] Favé, in *Des nouvelles carabines*, also argued for cavalry to become mounted rifles,[192] and the suggestion that regiments of yeomanry cavalry should be converted to mounted rifles was implemented in one or two cases in 1853.[193] Running counter to these trends was the view expressed by Sir Howard Douglas, that such expedients were all very well for specific colonial requirements but were too defensive in concept to be general cavalry doctrine.

Training in the use of firearms and in dismounted work gave to cavalry the opportunity to operate independently, to mount long-range thrusts in strength against the enemy's rear and communications. The validity of independent cavalry operations was accepted in Britain as applicable in the colonies and was embraced by Frenchmen in North Africa,[194] but in Britain at any rate it failed to grow into a general idea for the future employment of the arm. True, Jomini had spoken of 'some thousands' of cavalry acting in the enemy rear, and an advocate of similar schemes was to be found in the *United Service Journal* in 1831,[195] but no official brigade evolutions appeared to sanction or encourage such theories. Nonetheless, even that confirmed infantryman William Napier, writing in 1855, argued that the Crimea should have been predominantly a cavalry war. He talked of 20,000 cavalry ranging over the peninsula and cutting Sebastopol's supply-line.[196] Cardigan's ill-managed reconnaissance to the Danube suggested at least the possibilities of searching the open spaces in brigade strength.

One immediate limitation to these sorts of schemes was the problem of supply. It was, after all, to be the biggest constraint in the independent use of cavalry in the South African War: inadequate forage meant sore backs and broken mounts. But the far bigger constraint was the difficulty of producing a coherent doctrine in the circumstances of imperial campaigning. Against the firepower of the Boers, the value of cavalry had to be strategic, not tactical. It had to rest on the mobility of the horse away from the battlefield itself. Against poorly armed Zulus or Dervishes the cavalry would still have the opportunity to charge. Recognising the end of the *arme blanche* was difficult enough in the context of Europe alone; after all, the growth of firepower was gradual, as it progressed from rifle to breech-loader, from magazine-fire to machine gun. *Grande guerre* orthodoxy was itself unclear. Colonial precepts fed both sides of the debate. But at the least, they made for innovation as much as for atrophy. They also provided a keen edge of recent, and constantly updated, experience.

# 4

# Artillery: the arm of the future

## (i) The organisation of the regiment

In the second half of the eighteenth century Frederick the Great in Prussia and Gribeauval and du Teil in France had brought to field artillery a degree of mobility and standardisation that had transformed its potential on the battlefield. Their work enabled the guns to be concentrated rapidly on the decisive point and thus artillery became the third – and arguably the most powerful – arm.

Although Britain was a latecomer to these developments – the Royal Horse Artillery, for example, was not formed until 1793 – the equipment of her field artillery was able to profit from the experiments and mistakes of her neighbours. She was slow too to acquire sufficient quantities of guns, and only at Waterloo was her field artillery anything like the desired strength. The high-point reached by 1815 suggested a rosy exterior. But in truth the Royal Artillery had a rotten streak that had been only thinly covered.

In 1819, Lieutenant T. F. Simmons, R.A., published anonymously a pamphlet that pinpointed most of the problems that were to obsess gunners in the ensuing thirty years. In particular, on his first page, he identified the promotion system of the corps as one of its major deficiencies. In the cavalry and infantry purchase helped the current of promotion. But in the artillery and engineers promotion was by seniority, and peacetime contraction slowed the flow to a trickle. Simmons reckoned that by 1834 all the artillery's lieutenant-colonels would be aged sixty-three, and even the first lieutenants forty-three.[1]

The mechanics of promotion in the Royal Artillery, and their attendant problems, could not but stultify merit and discourage initiative. The Ordnance had its own school, the Royal Military Academy at Woolwich. All the officers of the Royal Artillery and Royal Engineers therefore received extended professional instruction: the same could not be said of the cavalry and infantry. But the superior training of Ordnance officers was not rewarded. They rarely received command and staff appointments, partly because they were not considered qualified to lead infantry or cavalry,[2] but largely because seniority promotion failed to produce men of the required rank who were sufficiently young or fit. In 1848, Grey suggested to the Queen that it would be politic to choose an officer from one of the Ordnance regiments for the governorship of Gibraltar, but admitted that selection was difficult as most of the possible candidates were too decrepit.[3] To achieve Grey's aim, Sir Robert Gardiner, a gunner, was appointed, although Anglesey, the Master General, thought him 'a silly vain man'.[4]

The top of the regiment therefore lacked distinction. Sir Hew Ross, the Deputy Adjutant General of the Artillery, and the Lieutenant General of the Ordnance during the Crimean War, was one of the senior artillery officers between 1840 and 1855. He consistently maintained such a bland image, even in the face of frantic demands for a reorganisation of the regiment, that, while C. W. Pasley could typecast him as 'an advocate of the present system', Colonel Brereton assured Pasley that Ross was simply its official rather than its real supporter.[5] Gardiner apart, not one general officer fully identified himself with the reform of the artillery, and only one colonel – F. R. Chesney. The real strength of the regiment in the late 1840s lay in its captains, including four – J. H. Lefroy, F. M. Eardley-Wilmot, E. M. Boxer and C. J. B. Riddell – who shared with Chesney the distinction of becoming Fellows of the Royal Society.

In 1838, Vivian, as Master General, suggested one possible solution to the incubus weighing on the regiment. Although virtually powerless to affect the current of promotion, he was at least able to fill the key posts at Woolwich with 'the ablest & most liberal minded of our Officers ... Men who will look about them to suggest & adopt impovements whenever they find this can be done with advantage – instead of resting satisfied to go on in the old jog trot way, & by & by allow us to be found half-asleep, whilst other nations, having had their eyes open, have stolen a march upon us'.[6] It was through the occupancy of appointments at the Royal Laboratory and elsewhere that men like Boxer were thus able to compensate for the inertia higher in the regiment. But their positions gave them little executive authority and their success depended on the favour of the Master General.

Furthermore, even if the right personalities had emerged at the top, fundamental structural weaknesses in the regimental command would have hampered a coherent expression of professional views. Between 1830 and 1854 no gunner officer had major executive powers. During that period the post of Lieutenant General of the Ordnance was suspended, and without him there was no gunner representation on the Board of Ordnance. The two senior officers of the regiment were the Deputy Adjutant General of the Royal Artillery and the Director General of Artillery.

Except for the years 1838 to 1843, when the two posts were combined, these functionaries were independent. The Deputy Adjutant General was the executive instrument of the Master General of the Ordnance in the control of the regiment's personnel and of its movements. Sir Alexander Dickson, who had commanded the artillery in the Peninsula, held the appointment from 1827 to 1840 and demonstrated that an experienced officer with well-formed opinions could secure the ear of the Master General. But for a weaker character the foundations of his powers were built on sand. The Director General of Artillery was in charge of the condition and types of guns, and was also responsible for the vitally important Field Train Department. This was concerned with the supply of ammunition and *matériel*, but in peacetime the department was reduced to a civilian staff of four commissaries, who could do little to decide on the likely expenditure of rounds per gun or to check on the quality and scope of the stores and depots.[7] In 1843 an Assistant to the Director General was appointed. Among other things, the Assistant was to act as secretary to the Select Committee of Artillery Officers, tellingly described as a Committee 'of opinion, not a Committee of decision'.[8] The Select Committee could not initiate discussion, and therefore, whatever its scientific worth, it was only liable to be influential when its aims coincided with that of its President, the Director General. Well might the gunners look wistfully across to the Royal Engineers, who had in the Inspector General of Fortifications a true leader, in charge of all aspects of that corps.

The commandant of the Royal Artillery was therefore the Master General of the Ordnance, a military man certainly but a political appointee, encumbered with the numerous other duties of the department, and never once a gunner officer. The range of his responsibilities, from the arming of the navy to the supply of accoutrements, from the manufacture of gunpowder to the training of cadets, meant that in any case his power for good was circumscribed. It was therefore essential that he should listen to the advice of his two senior artillery officers, the Deputy Adjutant General and the Director General.

From 1818 to 1827, the period of the post-war reductions, the Master General was Wellington. Quite apart from his inveterate inability to consult others, Wellington brought to the office his eighteenth-century derogation of the artillery as an arm. His public slights of gunners were interpreted as evidence of a deep-seated lack of understanding. The Royal Artillery rapidly learnt to temper its respect of the Duke. An officer, sent to Lisbon in 1827, denounced Wellington in a letter to the *Morning Herald* as having 'purposely rendered the British Artillery Corps in Portugal unfit for any service'.[9] The Duke had failed to grasp that Gribeauval and Napoleon had made field artillery potentially the most important part of an army in

European warfare. He continued to see the guns as no more than support weapons, and did not appreciate that sophisticated equipment was leading to greater specialisation. Peacetime economies and colonial garrisoning dominated his thoughts as Master General; they therefore dictated the pattern of regimental organisation to the detriment of its best tactical use, and even of its discipline.

The prime unit in the Royal Artillery was the company, and although the battalion existed on paper, it never formed or fought as such and its field officers did not hold active rank. In 1818 Sir Augustus Frazer published a pamphlet drawing attention to the difficulty of securing uniformity amidst so many different companies, each serving independently. He wished to see each battalion become an integral unit for service. Gunner officers switched companies so regularly that little chance existed for them to get to know their men, and for *esprit de corps* to develop.[10] In 1842, Colonel Firebrace suggested that each battalion should be allocated a group of colonial stations – one at Barbados for the West Indies, one at Malta for the Mediterranean, and two more for the Cape and Canada.[11] Thus each battalion could rotate between home, tropical and temperate climates, in the same way as an infantry battalion.

Colonel F. R. Chesney was an ardent advocate of an effective battalion structure for the artillery. Chesney agreed with Frazer's conclusions and added the rider that the virtual doubling of the number of battalions to twenty-four would relieve the promotion congestion at field officer rank, and thus confer two boons simultaneously on the regiment.[12] In 1840 he laid his scheme before the Duke of Wellington as chairman of the commission on promotion. But Wellington returned Chesney's papers, caustically observing that he was neither the secretary of the commission nor (any longer) the Master General of the Ordnance.[13] It is said that Wellington had considered the arrangement in 1823 and dismissed it in the light of the regiment's very small establishment and the delay and uncertainty in communication between the various colonies.[14] Sir Hew Ross challenged Chesney's case in his evidence to the 1849 select committee on army and Ordnance expenditure. Ross contended that most artillery service was done by detachment and that the company might therefore as well communicate direct with Woolwich as with some battalion headquarters in a colonial centre.[15] Ross totally ignored the small size of battalion that Chesney was advocating, about 400 men. An enlarged company that was fully self-sufficient and had its own field officers would have been a compromise that would have been adaptable to detached service.

Therefore the company structure was left unreformed. The system preserved the homogeneous character of the regiment, allowing an individual company to transfer easily from garrison to field duties and back. Without it, Colonel Lewis foresaw, 'special Corps of Artillery must be organized for special duties which would impair the general efficiency of every branch, taking into consideration the necessary employment of a large portion on foreign and colonial duty'.[16] This aversion to specialisation killed the suggestion made by Captain Riddell in 1852, and adopted by the regiment after the Crimea, that each battery should be eight guns, commanded by a field officer, divisible into two troops or companies and capable of being formed into brigades if necessary.[17] Riddell's organisation was appropriate for field artillery, but not for garrison artillery.

However, in preserving the unity of field and garrison artillery, the Ordnance, and Wellington in particular, failed to recognise that in fact the duties were distinct and fast becoming irreconcilable. Particularly since Gribeauval's day, field artillery had to move lighter guns at speed, to understand and control horses, and to be fit and have an eye for country. By contrast garrison artillery needed to comprehend the complexities of a different range of equipment and were confined to manhandling heavier pieces in a stationary position. The foot artilleryman in the British army was expected to combine all these skills, to be proficient in infantry drill, garrison, siege and field artillery, and stable duties. Any one of these he might neglect for years, and in the interim not only forget but lose the fitness of youth that had enabled him to do it. Consequently Chesney and other writers called for the division of the regiment into three distinct branches, horse, field, and garrison artillery. In 1840 Dickson considered the proposition, but only in so far as it would provide garrison artillery for home service, and so not surprisingly concluded that it would throw an inequitable burden of foreign service on the rest of the regiment.[18] However, the colonies stood as much in need of heavy artillery for the protection of their coasts, harbours and forts as did the mother country, and either a separate garrison artillery regiment with its own rotation or an integrated battalion divisible into field and garrison companies was mooted. Neither scheme was likely to find approval when the small size of the artillery force in any one colony compelled all of it to be fit for both garrison and field duties.

Nonetheless in those places where the area of operations was restricted, an island or a trading-post such as Hong Kong or St Helena, force was given to the oft-repeated demands that infantry be taught 'great gun' drill. Simultaneously and more coherently, a policy began to emerge of employing locally raised companies of garrison artillery. In 1842, volunteers from the West India Regiments were formed into a company of African Artillery for service in Jamaica,[19] and in 1846 approval was also given for gun Lascars, Malays from the Ceylon Rifles or Indians from the East India

Company's service, to form an artillery force at Hong Kong.[20] But although these troops were undoubtedly garrison artillery in a way that none of the rest of the regiment was, it would be erroneous to construe that their employment constituted an acceptance of garrison artillery as an ideal. The companies of Royal Artillery whom they replaced remained in the colonies, the purpose being to enable the Europeans to be quartered in a more salubrious climate. Africans, who had not been trained at Woolwich, could not, it was felt, have pretensions to gunnery expertise, and therefore, as soon as danger threatened, the Europeans were to be brought from the hills to replace them.[21] The application of colonial garrison artillery remained no more than implicit.

The decision not to divide field and garrison artillery was compounded in its consequences by Wellington's introduction of the gunner-driver system. Before 1822, the Corps of Drivers had been separate from the rest of the regiment. Their bad reputation provided a plausible excuse in 1821 for instructing all gunners to train also as drivers, and in 1822 the Corps of Drivers was absorbed into the regiment as a whole. Such an arrangement enabled the regiment to be as flexible as possible in colonies where the detachments' strength might be low and specialist replacements unavailable. Dickson maintained that under the gunner-driver system 5,000 men were equivalent to 7,000 under the old arrangement of separate duties;[22] it brought the reduction of the Royal Artillery in 1823 from 62.5 per cent of its 1815 strength to 72.4 per cent (as opposed to a 62.1 per cent reduction for the rest of the army) without theoretically lowering its battle strength.[23]

Economically it was sound, and of course bore the Wellingtonian stamp of pragmatism. Militarily it exacerbated the confusion already prevalent over the duties of field and garrison artillery. Large and strong gunners were required to manipulate guns, whether in the field or in garrison, and therefore the minimum height for enlistment in the regiment was 5 ft 8 in. Drivers, on the other hand, might reasonably have been smaller in order to limit the burden on the horse, particularly in view of the development of faster field artillery. Both gunners and drivers took legitimate pride in the different skills and techniques of their jobs. A strong lobby therefore called for the separate enlistment of artillery drivers, and there were good grounds for its optimism since the 1819 committee, formed to review the peace establishment of the regiment, had concluded that in war drivers would have to be a separate body.[24] In 1847, Ross acknowledged that the subject was a controversial one among gunner officers, 'many ... having thought that without a separate body of Drivers, no Artillery could be made thoroughly efficient in the field'. In reply, the Master General, Anglesey, acknowledged the advantages of a separate corps of drivers, but was constrained by the expense.[25]

Fig. 16 *Royal Horse Artillery: 'Fire' ('Right Flank thrown back for Action, Right')*. Lithograph, drawn by G. B. Campion, 1846 (National Army Museum)

Thus the gunner-driver system survived, Dickson unwittingly conferring on it the blessing of his Peninsular experience and Ross being too weak in character to gainsay such authority. Vivian confessed that, as a cavalryman, he had expected the system to fail but had his doubts dispelled while Master General.[26] Within the battery a consistent allocation of duties by the adjutant gave specialist training to those best suited to the respective skills of driving and gunnery. Because no major upsets occurred in Portugal in 1827, in Canada or at the Cape, it was concluded – without any direct evidence – that the system worked in the field.[27] But in 1854 the field artillery was short of trained drivers, and disaster was only averted by the cessation of field operations after Inkerman. The gunner-driver system was eventually abolished on Ross's retirement in 1858.

That the emphasis on the unification of duties within one man, or roles within one regiment, did not lead to the greatest efficiency was shown by the Royal Horse Artillery. It was a specialist arm, with its own scale of equipment and with its men (although not its officers) permanently attached to it. Even its drivers were expressly enlisted as drivers.[28] The result was that the Horse Artillery was 'an example and a school for our

Field Artillery. It is an example', Sir Robert Gardiner wrote, 'in its system and discipline; it is a school from which the Officers of our Brigade Artillery derive their knowledge, and judgment, and experience, in the care of horses, and that thorough knowledge of stable duties, as necessary to Officers of Artillery in command of troops and brigades, as that of gunnery itself.'[29] The 1849 committee on army and Ordnance expenditure, urged to dispense with the cost of horse artillery in peacetime, concluded that it was not merely the lengthy training that justified its upkeep but also the fact that 'it is to be regarded as the school of the whole regiment for field artillery practice'.[30]

Nonetheless hard-line reformers, such as the *Naval and Military Gazette* and the *United Service Gazette*, found the Horse Artillery 'a leprosy ... contaminating ... the whole corps',[31] 'consuming the very vitals of our Field Artillery'.[32] Much of the abuse was directed not at horse artillery *per se*, but at the British system of organising it. As a school for field-artillery officers it was not as efficient as it might have been.

The Royal Horse Artillery was an élite, and appointments in it were therefore highly prized. One calculation reckoned that only a sixth of the regiment's serving officers had passed through it.[33] Although it was declared that all appointments to it were to be granted impartially and in rotation, according to service and standing in the corps, a vacant captaincy in 1846 was alleged to have gone to a lieutenant already in the horse artillery and not to the most senior eligible captain of the regiment.[34] Part of the reason for the competition for places was that India, the only foreign station where British cavalry normally served, had its own indigenous horse artillery: therefore the Royal Horse Artillery did not go abroad. In 1852, it was calculated that the Royal Horse Artillery, with its forty-six officers for eighteen guns, had cost £1½ million since 1818 with no return in service. Colonel Firebrace jumped to the easy conclusion that it 'appears to be kept up for the sole purpose of exhibiting after the fashion of *saltimbanques* and rope-dancers, before foreigners of distinction'.[35] The manoeuvres became increasingly rapid for the purpose of show, and at a review in 1853 a gun was fired while the number two in the crew was still ramming home.[36]

The Royal Horse Artillery showed by its fitness in the Crimea that these types of criticism sprang more from jealousy than fact. The main case for disbanding the Horse Artillery in peacetime revolved not round its alleged foibles, but around its proportionately larger claims on the niggardly supply of horses vouchsafed by a miserly government. In 1815, the Royal Artillery had 14,000 horses; in 1831 the regiment had only 600, a reduction far greater than that in the number of men.[37] Furthermore, the horses at Woolwich were employed as much on labouring duties at the

Royal Arsenal as on specific artillery instruction, and many of them were said to be unfit for the draught of guns. Against this background, it was calculated that each horse-artillery gun required fifteen men and fifteen horses to bring it and forty-six rounds (the total number carried in the limber) into action, whereas a normal field-artillery piece (where the crew in theory marched on foot) required thirteen men and eight horses to move it and 194 rounds.[38] Therefore, for a loss of the six troops of horse artillery (eighteen guns), the same stock of horses could be variously divided to give twelve efficient field batteries (seventy-two guns) or twenty-seven four-gun batteries (108 guns).[39]

The cuts in manpower at the end of the war, although not as severe as those in horses, were nonetheless savage. The Royal Artillery dwindled from 25,000 men to 6,000.[40] In 1815, Sir Robert Gardiner wrote in 1848, 'an arm of the service was annihilated'.[41] As early as 1828, the Secretary at War, Hardinge, although then reducing army estimates in several other directions, could bemoan that 'Britain alone [in Europe] is so situated that not an artilleryman could, for the first year of war, be spared for foreign service.'[42]

Lack of men incapacitated not merely the field artillery, already in any case hamstrung by insufficient horses, but also those on garrison duties. In 1828 the increased colonial commitments of the British army meant that artillery detachments were to be found in Gibraltar (400 men), Malta (171), the Ionian Islands (233), the Cape (83), Ceylon (133), Mauritius (78), the West Indies (294), Jamaica (213), North America (648) and Bermuda (14). By 1847, not only had the commitment to the Canadian colonies increased to 998, but also St Helena, Hong Kong, New Zealand and Hudson Bay had been added.[43] In 1843, Sir George Murray reported that colonial reliefs had become so irregular and so far apart that the discipline of the regiment was threatened. Wellington had aimed as Master General to have five companies in reserve at Woolwich to be used for reliefs or as a disposable force to garrison an undermanned and threatened colony.[44] In 1838, it was possible to maintain only three of the five.[45]

Furthermore, despite economies in men and the postponement of reliefs, no individual colony was sufficiently garrisoned. 'It is an unquestionable fact', Murray asserted, 'that the amount of the Artillery force at every one of the stations where it is at present employed, is much below that to which it would become necessary immediately to raise it in the event of war.'[46] Too much reliance on a disposable artillery force at home would have been unwise in the days of slow communications, and the neglect of some stations in order to build up others would have left little flexibility to cope with contingencies. Thus, in 1838, the Cape, midway between two Kaffir Wars, had under 100 gunners, or only two men per gun.[47] In 1843, promptings from the artillery commander in South Africa, Captain Eardley-Wilmot, did at least ensure a second company in time for the later Kaffir Wars, but the disposable field artillery in 1851 still only comprised two 12-pounder howitzers and three 6-pounder guns.[48] In 1848, 4,812 guns were mounted in garrisons throughout the empire, but only 9,000 men were available to operate them: in 1847, the Ionian islands, Gibraltar, the West Indies and Malta had about one man for every two guns.[49] Gardiner reckoned that the Gibraltar garrison required ten, not five companies, the Cape four, not two, Corfu six, not three, and similar increases were called for in North America, Malta and the West Indies.[50]

But if the problem of manning garrison artillery in the colonies gave rise to concern, the situation at home seemed potentially catastrophic. With the fear of French invasion growing throughout the 1840s, the artillery of Britain's coastal defences was again examined. In 1843, of the main south coast ports, Dover had ninety-nine gunners, Portsmouth sixty-nine, Devonport seventy-five, Chatham and Sheerness one, Pembroke thirteen, Landguard four and Harwich eight. Moreover the Channel Islands, a possible staging post for a French invasion, could only boast a further seventy-nine.[51] These figures referred only to men and reflected inadequately the low numbers of guns themselves. A refurbishing of coastal defences was envisaged and at the beginning of 1854 it was calculated that in the forts, towers, castles and batteries of Britain, excluding the Channel Islands and Ireland, 1,664 guns were either in place or scheduled to be mounted and that they would require 13,038 men, or virtually the equivalent of the then-strength of the entire regiment.[52]

An alternative to this heavy expenditure in bricks and mortar, guns and men, was the establishment of efficient movable batteries, based on central secure positions, that could be rushed to wherever the enemy might land. However, the field artillery of Britain not merely was rendered inefficient by the gunner-driver system and the Wellingtonian failure to understand its correct use, but also suffered more than either the garrison or horse artillery from the deficiencies of men and horses.

The 1819 committee stipulated an allocation on service of one battery of six guns per 4,000 men.[53] Simmons's pamphlet of the same year suggested twenty-two guns per 800 men. Slender though the 1819 requirement was, it was infinitely better than the actual figure of forty field guns horsed in 1850 (excluding horse artillery), as against 500 in France (or 3,759 without horses), 492 in Prussia, 1,020 (or 400 fit for the field) in Russia, and even eighty-four in Belgium.[54] No attempt had been made to implement Dickson's 1832 proposal of a total field artillery force of 246 guns, of which eighty-six (excluding very light horse artillery) were intended specifically for home service.[55]

The actual strength of the field artillery at the beginning of 1848 comprised five two-gun troops of horse artillery without ammunition waggons in England, and two four-gun troops in Ireland. The companies stationed at Woolwich were formed for instruction into four four-gun batteries, each with forty-five horses. Four batteries in Ireland were on a similar footing; one troop of two guns at Leith was horsed but two more at Chester and Weedon were not.[56] Gardiner concluded that even this nominal strength was misleading; two of the Woolwich batteries were 'so little advanced in their necessary drill and training as to be quite non-effective for the purposes of service'. He concluded that fourteen guns could be sent to the coast from Woolwich, provided that horses were taken from two brigades under instruction to draw the ammunition waggons.[57] The following year, even Sir Hew Ross confessed that of the eighty guns normally then available in the United Kingdom and Ireland none could be deemed fully equipped, all of them being short of horses.[58]

The principal requirement to enable the Royal Artillery to become efficient was money. With an increase in the Ordnance estimates, men and, more importantly, horses could become the raw material from which proper field and garrison artillery might be forged. Colonial defence, the cause espoused by Vivian as Master General, was not the key to the Commons' coffers. Glenelg, as Secretary of State, refused an increase of more than three companies in 1838.[59] This augmentation, and that of the following year, took the form of an additional 576 men for the existing companies – an increase that could only be deemed generous when contrasted with the existing total of a paltry 6,071.[60]

The fear of French steamer-borne invasion and the consequent concern for home defence were to become the saving of the regiment. Vivian's scepticism about the danger to Britain's coast-line was overthrown by Wellington. With the return of Sir Robert Peel's government in 1841 and Murray's appointment to the Ordnance, Wellington had two powerful allies who shared his concern.

In August 1844, Murray asked Peel if he could recruit to establishment, that was to say 200 above the current strength. The Channel Islands, Murray pointed out, had only one company, and training at Woolwich was having to be curtailed to meet the demands even of peacetime service.[61] In an 1845 memorandum to Murray and Peel, Wellington spoke at length of his proposals for the defence of the south coast, and suggested an increase of at least 500 men for the batteries he proposed.[62] Not until January 1846 was Peel won over. Murray asked for 'the most limited augmentation' of 720 men, or ten per company,[63] and in the event the Prime Minister added to the 720 a tenth battalion, giving a total increase of 1,542 men in March 1846. Peel's fall made no difference to the government's new policy, Anglesey coming to the Ordnance as impressed with the United Kingdom's exposed state as his predecessor, and Grey acknowledging that an increase to the Royal Artillery must take priority over one for the army as its training was longer and more complex. Thus in 1847 the strength of the regiment was augmented by an additional twenty companies and in 1848 two new battalions – the 11th and 12th – were added to help promotion in the higher ranks. In 1846, the strength of all ranks had totalled 6,709; in 1848, it stood at 11,347.[64]

The work of Murray and Anglesey belonged to the orthodoxy of static defence. It simply provided gunners for the coastal fortifications, but it did not solve the problems of the field artillery or recognise the role it might play in a more flexible United Kingdom field army. Each troop of horse artillery in Britain had fifteen horses added to its establishment in 1848, so that it might take four, not two guns into the field, and a further ninety horses were sent to Woolwich to enable two more field batteries to commence instruction.[65] Beyond this, however, the 1846–9 increases exacerbated the prevailing weaknesses of the Woolwich field battery instruction, providing more men for proportionately fewer horses.

Thus in 1851 the total strength of the field artillery in the United Kingdom was still a mere forty guns, together with the additional twenty-eight of the horse artillery. Woolwich itself could not provide the seventeen guns required for Wellington's funeral the following October, and Sir Hew Ross advised that only four horses (rather than the proper draught of six) could be allocated to each if 'a good turn-out' was to be attained.[66] At Chobham camp in 1853, a fourth battery had to be fleeced to complete the equipment of the three sent, two of which in any case had only enough horses to employ the light 6-pounder, rather than 9-pounder, guns.[67]

The drive to render the field artillery fit, and in particular to secure for it more horses, was two-pronged. The *Naval and Military Gazette*, employing Sir Charles Napier's friend, Captain F. Robertson, a former gunner and an enemy of the horse artillery, took a most vigorous line on the subject. At the beginning of 1848 it sought an immediate increase of five battalions and a minimum field artillery of fifty guns, with a long-term objective of thirty-six batteries for England (216 guns) and eighteen for Ireland. Horses were the great desideratum and the *Gazette* espoused Pasley's call for 3,000 for the field artillery. The other prong had a sharper point in the shape of an artillery general. In a lengthy memorandum of January 1848, Sir Robert Gardiner told the Prime Minister, Lord John Russell, that he wanted eighty field guns 'constantly horsed, equipped and ready for service' in the United Kingdom, and suggested that they should be formed in eight-gun batteries of 9-pounders. Within two weeks, Gardiner had revised his estimate upwards, to 100 or 120 guns.[68] Sir

Robert won the support of Prince Albert, but Russell was not sympathetic. In March 1848, the general published his views in a pamphlet addressed to the committee on army and Ordnance expenditure, calling for an immediate augmentation of three troops of horse artillery, thirty companies of 100 gunner-drivers each, and 1,000 horses. Thirty-six more field guns would thus be made effective, and the strength of the regiment would be increased to 14,550 other ranks and 2,022 horses.[69]

In reply to all this vociferous pleading, Anglesey was – by the lights of his predecessors – successful. To the two extra field batteries and the doubling of the horse artillery troops in England (165 additional horses) which he had effected in 1848, he secured at the beginning of February 1852 another 1,000 men and 135 horses.[70] However, within days of Grey's approval, Anglesey had been replaced as Master General by Hardinge. The First Sikh War had left the former Governor-General of India with a very clear idea of the efficacy of large-calibre field artillery, well employed, as it had been at Ferozeshah and Sobraon, under the direction of French officers. Hardinge believed that 'A powerful artillery may make an opponent hesitate or if he were to attempt invasion, our state of preparation would afford us the best means of defeating it.'[71]

On 18 May 1852, Hardinge drafted a memorandum calling for 120 guns to be ready for the field by the end of the year: the Treasury approved. Four months later he produced a second paper that was possibly the most comprehensive and important survey produced on the requirements of a field army between 1815 and 1854. 'The permanent defensive materials required for a large field force of all arms, must be placed in time of peace in a more perfect footing than they now are', he warned. 'Postponements, hazarded for financial convenience from year to year entail the chances of more misfortune and disgrace than in other human affairs, and must not be left to the Chapter of accidents.'[72]

At the centre of his argument lay the artillery. It must be proportionately larger in peacetime, particularly since it would be the arm an invader would lack on first landing. Hardinge therefore planned on 244 guns being ready by the end of 1853, and had, as his eventual target, the total – immense in the context of the previous decades – of 300 field guns for home service, 180 of them for England and Wales. Even this he deemed the absolute minimum, since with the auxiliary forces it would give a ratio of only one gun for every 600 men. A hundred extra workmen had been enrolled at the Ordnance's Carriage Department to progress the work, and so strong was the Master General's impatience that he advocated recourse to outside contractors. No detail escaped his attention. Three hundred guns would require, he computed, 800 carriages, 3,232 wheels and 200 ammunition waggons. Each gun should have 166 rounds in action and eighty-four in reserve. Four thousand horses would be wanted for the guns and 4,600 for the waggons. At the moment the artillery only had 1,300 horses: a further 1,000 must be obtained that year just to enable proper peacetime training.

Although transferred to the Horse Guards, Hardinge would not let his proposals wither. A third memorandum appeared in November 1852, setting a target of 200 9-pounders and horse artillery 6-pounders by the end of March 1853, and 300 field guns by the conclusion of that year. Six hundred rounds per gun, or 100,000 altogether, should be prepared. The existing store of 150 rounds per gun was deemed sufficient for the 1,000 garrison guns in England. To man this vast force, 7,000 gunners (made up of 3,000 regulars, 1,000 artillery pensioners and 3,000 militia artillery) should be left in the forts and 7,000 regulars (or an increase of 2,000 on the existing numbers) should operate the field guns.[73] A report by Raglan showed how far and how fast Hardinge was expecting the country to catch up with its neglect: in November 1852 still only ninety guns (fifty-eight in Britain and thirty in Ireland) were ready to take the field, a further 116 pieces were not horsed, and of these 206 'the greater part [were] of low calibre on a peace establishment & many of the carriages old & not in a condition to go through a campaign'.[74]

Probably thanks to the Prince Consort's support, a supplementary Ordnance estimate was approved in December 1852, granting 2,018 men and 1,000 horses.[75] According to the Prince's secretary, Charles Grey, this would give 200 field guns, the target of 300 being postponed.[76] However, in a letter to Newcastle of January 1853, Hardinge clearly expected the target of 300 still to be approved in the course of the year, 200 being the interim target for 'April next'.[77] The implication, therefore, is that the timing alone was being modified, and that 1854, not 1853, was to be the date of completion. Clearly the reports of the artillery's inadequacies at Chobham would confirm such a construction. The regiment's historian says that by April 1853 only 128 guns were in service with the field artillery,[78] but a French report in July of that year confirmed that the British were preparing 300.[79] From the end of 1852, Raglan, Hardinge's successor, set about increasing both the horse and field artillery to six-gun batteries and put the emphasis, as Hardinge had desired, on arming them with 9-, not 6-, pounders.[80]

Inevitably the full impact of Hardinge's work could not have percolated throughout the regiment by the outbreak of the Crimean War, but to him, more than any other, must belong the credit for Burgoyne's report that the Royal Artillery in the East had shown itself 'very efficient and very good'.[81] Hardinge had not envisaged the employment of his creation in an expeditionary force, but it is sufficient indication of his talents that what had been intended for the Sussex Downs should have proved so adaptable to the uplands of Sebastopol.

### (ii) Training and trials

Implicit in all the arguments about numbers of men and horses, and about the gunner-driver system and specialisation within the regiment, was the assumption that the existing system thwarted any adequate training. Hardinge's opening premise in his plea for an enlarged artillery affirms this: artillery, he said, 'is the most difficult arm to create: it requires time to drill the men to ride and drive and to train the horses'.[82]

However, sufficient time was not granted. In 1827, four field batteries were established at Woolwich, three of which were for the purposes of providing the nation with properly trained field artillery, and the seventy-two companies served with them in rotation. In 1834, it was calculated that such a ration would give each company a bout of field instruction once every eighteen years.[83] Nor was the recruit's course sufficiently long to compensate: by 1849 the minimum of eighteen months necessary to teach a new gunner had dwindled to at most eight months and at times allegedly three.[84]

Before 1846, it was always the intention to keep twenty-six companies at Woolwich for instruction, garrison duties and colonial reliefs; in practice there were twenty-four. In 1849, the number was constant, although now the establishment of the regiment had jumped from seventy-two to ninety-six companies. Therefore the bottle-neck at headquarters was proportionately narrower, especially since a much greater number than usual were recruits.[85] One of the four field batteries had always been employed on public works at the Royal Arsenal, and a difficult situation was aggravated in 1847 when a second was withdrawn for the same purpose.

Eight field batteries would have brought the Woolwich course back up to twelve months, and ideally, of course, as many should have been created as would have ensured that every company was never more than two or three years away from battery duty. The lack of horses was the great limiting factor. The ninety secured by Anglesey in 1848 mounted two more field batteries at Woolwich, and in 1849 Ross reported that, although a further two (to make eight) were still required, the 'instruction of the regiment ... had been of late improved'.[86] Hardinge's augmentations made the batteries already at Woolwich efficient or added new ones elsewhere, but he did not increase the numbers of batteries of instruction. However, in 1854, the construction of a railway within the Royal Arsenal meant that forty rather than 106 horses were required for labouring duties, and the balance could therefore be released back to the batteries.[87]

The demands of the Arsenal drew off men as well as horses. One angry correspondent claimed that the companies at Woolwich spent three-quarters of their time doing fatigues or working in the docks.[88] Although this was an exaggeration, the penury of the public was again prejudicing the potential efficiency of the artillery. Before the 1846 augmentations gunners at Woolwich spent every third day on guard, and mundane garrison requirements therefore interfered with the recruit's already too brief instruction. Furthermore, even when the field exercise was done, the need for the batteries to appear on ceremonial occasions put a premium on speed and dash. The drill therefore rarely took into account the contingencies which in action might impede the operations, and line and elevation were never given, so that the laying of the gun was not properly practised. One of the worst developments of the desire for showmanship was the mounting of gunners, who would otherwise proceed on foot, on the gun limbers. In peacetime manoeuvres, when ammunition was not carried, this was possible, but it bore no relation to wartime requirements, as the weight of the rounds would preclude anything else from being transported on the limber.[89] Even at Chobham, the field batteries exercised without ammunition waggons. The surprise therefore cannot have been great when it transpired that the detachment sent to the Cape in 1847 could not harness their horses[90] or when the battery ordered from Woolwich to London on 10 April 1848 was found inadequate even to face Chartists. Much of the problem was that the officers, not being permanently attached to specific companies, took little interest in the battery instruction and instead left it to the N.C.O.s. Of the four batteries at Woolwich in January 1849, Sir Robert Gardiner claimed that 'not one has ever been out under its Officers!'.[91]

However, beginning in that year, the combined assaults of Gardiner and the *Naval and Military Gazette* bore some fruit. The spate of augmentations had ceased and the number of field batteries had increased: the basis for improvement had therefore been laid. In addition the superintendent of field-battery instruction, Lieutenant-Colonel Anderson, emphasised the duties of artillery in the field, and the time devoted to infantry or carbine drill was reduced. The gunners were forced to march on foot. Eardley-Wilmot, who the previous year called his regiment 'little better than a militia corps', reported to Lefroy in June 1849 that 'The regiment generally is looking up in many ways, and a good deal is being done in the garrison by more judicious arrangements as to duty and employment.'[92]

The 1846–9 augmentations speeded the rotation of companies and so brought them to Woolwich for field instruction more frequently. But at all

Fig. 17 *Royal Artillery Repository Exercises, 1844*: instruction in use of siege artillery at Woolwich. Engraved and drawn by John Grant for *The Army and Navy Register and Woolwich Gazette* (National Army Museum)

other stations, with the exception of those in Ireland and Canada, battery work was totally neglected. During his absence from Woolwich, the gunner might do little more than fire a few salutes. Too often in garrison his time was devoted to sentry-go, and other infantry duties. Provision was therefore required for proper artillery instruction in the main gunner stations. In his memorandum of September 1852, Hardinge urged that batteries for field instruction be set up at Portsmouth, Plymouth, Sheerness and Dover, for 'if nothing be done, 1500 able artillery men are left in garrison, unpractised to take the field, for want of instruction in driving, although they have passed through the Woolwich course of instruction'.[93] Field batteries were consequently distributed more evenly in the United Kingdom in the course of 1852 and 1853.

The traditional duties of garrison artillery were on the whole well taught at Woolwich. All companies in their time at headquarters underwent courses at the Laboratory – on the making-up and composition of ammunition – and at the repository. The latter covered the shifting and mounting of all types of guns, and the use of the equipment associated with them. Little doubt existed as to its excellent arrangement. 'I met with nothing on the Continent', Captain Spearman wrote in 1853, 'at all comparable to it, considered either as a depôt of arms, implements and models, illustrative of the art of war, or as a School of Practical Artillery, so far as the manoeuvres of the arm are concerned.'[94]

However, the strains on the instruction departments in 1846 to 1849, and before that the demands of the battalions abroad, meant that too little time was devoted to the repository, and in 1848 it was said to have been submerged altogether.[95] This problem resolved itself after 1849, but it threw a greater emphasis on the school established in the repository for selected N.C.O.s. Even this course became somewhat limited in duration during the late 1840s, but its contents were admirably summarised in a number of pocket books, Captain Griffiths' *The Artillerist's Manual* having the approval of the Master General and being kept cheap deliberately to attract N.C.O.s. By 1850, the N.C.O.s' school was a twelve-month course for only thirteen pupils, and thus, whatever the deficiencies of the training at headquarters for gunners between 1846 and 1849, qualified instructors were despatched to battalions throughout the colonies.[96]

However, proficiency in bringing the gun into action, whether in a fortress or in the field, was of little value if unaccompanied by experience in firing it or by practical knowledge of its capabilities at long ranges and with various types of ammunition. In 1836 the annual allowance of practice ammunition was thirty-two rounds per gun and sixteen rounds per mortar. Simmons in 1819 correctly foresaw that rigid economies would mean that the opportunities for practice would be curtailed. He therefore pointed out that, so as to derive the maximum benefit in instruction, each shot must be all the more carefully observed in its performance. The reverse was the case. 'The whole course of practical instruction', a gunner officer wrote,

> consists in the lazy expenditure of three rounds, morning and evening, during the months for exercise, from certain guns at fixed ranges – the charge, elevation, length of fuze, and so forth, being all prepared before hand, by a laboratory sergeant appointed for the purpose, and who furnishes it on paper to the party for exercise. There is thus nothing to induce either officers or men to think, or to force them to rely on, and therefore improve, their own resources.[97]

Until a reasonable supply of practice ammunition bred a certain confidence in the use of the gun, little would be achieved. In 1848, an increase to eighty rounds per gun, or forty per mortar, was sanctioned,[98] but even this could only be described as generous in the context of the previous deprivation. In September 1853, Raglan wrote 'We must look forward to a greater expenditure of Ammunition in training and keeping up the instruction of the Artillery men.'[99] In particular, no allowance of shrapnel was stipulated, and it was perhaps Raglan's minute that secured an increase to

140 rounds per gun, of which sixty were normally shrapnel.[100] The opportunity given for regular practice thus encouraged a spirit of emulation and competition between the rival gun crews, itself an explanation of the improvement noticed in some quarters by 1852.

Nowhere was the lack of proper practice more sorely felt than at Woolwich itself. The range on Plumstead marshes adjoined the Arsenal and was therefore conveniently situated for batteries undergoing instruction. But 'as the line of fire crosses the river, the practice is almost invariably interfered with by the passage of steamers and sailing vessels, and it is often stopped entirely by their anchorage within the range of the shot'.[101] In 1841, 92,492 vessels passed from London to below Woolwich, together with a further 50,000 or so smaller craft.[102] Eardley-Wilmot reckoned in 1849 that on three out of five occasions the gunners left the range without having fired, and Captain Grant thought five or six abortive visits normal.[103] The ground between the firing point and the river was also let for agricultural purposes, and claims for losses to Mr Wallis's livestock were not infrequent. Haymaking was an additional hazard. A fair example of Plumstead's problems can be gleaned from the efforts made to conduct trials on various types of percussion shell in 1841. In the first week, firing was suspended by two barges dropping anchor and refusing to move. Two weeks later, after an hour's wait for the river to clear, three rounds were fired, but then the remainder of the day was wasted waiting for a gap in the traffic to enable the other four to be discharged. In the third week, the first series of trials was completed, but further firings had to be suspended as it was now August and the hay crop was due to be brought in.[104]

Even when firing was possible, Plumstead did not provide a very good guide for novice artillerymen. The maximum range was twelve or fifteen hundred yards. Field artillery firing at eight hundred yards got a fairly realistic practice, as did the short-range lobbing fire of mortars and, to some extent, howitzers. But the capabilities of the heavier ordnance developed in the early 1840s remained unknown. A 32-pounder with only 4° elevation and a four-pound charge would always range over 3,000 yards, and even a 3-pounder shot was known to have reached 2,760 yards.[105] In practice, therefore, the piece was always undercharged, ricochet firing was impossible, and often experimental firing had to be performed at Dover or Deal.

Initially attempts were made to realign the range at Woolwich. Since it was situated on a bend in the Thames, and pointed north-east, the logical solution was to change the axis to west–east so that the firing would be parallel to the river and independent of shipping traffic. But the river still lay 100 yards beyond the butt, and in 1838 the Master General quite rightly vetoed the proposal. Instead he initiated a second report which favoured an almost due east orientation, which would give 6,496 yards clear until the line of fire next met the river. However, this plan affected 110 acres of farming land, and Vivian thought that the expense of £22,651 was unjustifiable on a range that even then would not be ideal. In 1842 Lord Bloomfield, who commanded at Woolwich, revived the whole question in an impassioned plea to Sir George Murray. However, the east–west range at Plumstead now involved twenty-nine landowners, and ideally the compulsory purchase of 966 acres, the reselling of 346, and the letting of the rest for nine months of the year when the range was not used. Since this would only give a range 600 yards wide, the danger from the deflection of heavy guns' shot was not ruled out, and an 1843 committee abandoned the east–west range in favour of a scheme similar to that of 1838. This would avoid the anchorage in Gallion's Reach, and, although inadequate for experiments, would suffice for instruction. At an estimated gross cost of £41,935, the Royal Artillery would get a range of 1,500 yards, with a further 870 yards between the butt and the river bank. Not surprisingly the Treasury refused.[106]

From the beginning of the decade it had been clear that for experimental purposes, at least, no amount of tinkering at Woolwich would suffice. The 1843 committee of Ordnance officers concluded that no hope existed of finding an adequate range at headquarters, and that a start should be made elsewhere. In May 1846, Lieutenant-Colonel Dansey, R.A., summarised a vitriolic report by asking 'Who would not prefer the creation of an enlarged and permanent establishment adequate to the pressing demands of the service and worthy of the nation to resorting to an inefficient and derogatory expedient.' Such a site should be long enough for the greatest known distances achieved by artillery, and wide enough for all deflections; all shots should be markable, all projectiles recoverable.[107]

In 1844, Sandwich flats had been examined with this in mind, and in 1845 a survey was done of Landguard Fort, near Harwich.[108] But Landguard Fort could only offer a range of 3,000 to 4,000 yards, and was far outweighed by the 1845 report of two officers, Dansey being one of them, on the merits of the sands at Shoeburyness. The tide was off the area under consideration for eighteen hours a day, shots could on the whole be easily observed, and vessels of sixty to eighty tons could steam from Woolwich in seven hours to discharge any equipment that might be required.[109] By the end of 1847 the first rockets had been fired at what is still the home of the regiment's experimental establishment.

However, the development of Shoeburyness was slow. More land was bought there in January 1849,[110] and not until June of that year did troops go down to construct temporary accommodation.[111] The following month the Lord of the Manor created problems, apparently over the threats to his fishing rights.[112] Although experimental firing was conducted at Shoeburyness in 1850 and 1851, at this stage the possibility of an instructional range at Woolwich was again mooted. The arrival of Hardinge at the Ordnance gave the Shoeburyness project a new impetus. Colonel Chalmer was instructed to see that the ground was 'sufficient for the longest range', since it must be 'the most perfect range in the United Kingdom'. Further land was hired with a view to its eventual purchase, and the following month fitting-out again began in earnest.

At the same time Hardinge grandly, but not inaccurately, forecast the future. 'We may consider this ground, will for the future be the Practice Ground for the greatest Ranges required for artillery practice: & that Detachts for instruction & Experiments will be sent from Woolwich every year – & every description of Gun should be mounted, ready for Experiment, information and if possible for improvement.' The result will be 'that every description of gun practice for the Land Service – & even for the Naval, may be brought to the test of constant experiment – & that year by year the result of these Experiments may be recorded by the Dir Genls orders in such a manner as may be most conducive to the improvement of the Artillery in scientific research, & practical efficiency'.[113]

The range was now said to be five miles of sand and the same extent of beach, and throughout the summer of 1853 two companies were stationed there, engaged in experimental firing. However in 1854, the regiment again flirted with Plumstead as a range for instruction, further land had to be acquired at Shoeburyness in 1855, and not until 1862 was its use as an artillery range guaranteed in an Act of Parliament.[114] Provision for proper practice by the artillery at headquarters had not been made by the outbreak of war, but Shoeburyness allowed co-ordinated and continuous trials and experiments in the use and development of ordnance from 1852 onwards. The challenge of the breech-loading rifle to the artillery's growing preponderance on the battlefield meant that it had come not a moment too soon.

### (iii) Field artillery

The lack of a proper experimental range and Wellington's failure to understand the specialised and increasingly important use of field artillery were symptoms rather than causes of the relative quiescence in artillery developments during 'the long peace'. In 1852, Major Hector Straith, in his definitive *Treatise on Fortification and Artillery*, observed that 'very little change has been made in the outward form of guns for the last 200 years'.[115] Yet technology, if not yet upsetting the order on which the smoothbore cannon was based, brought it, within its own limitations, considerably nearer perfection. Peace provided opportunities for more studied observation of the performance of shot and shell, and the tendency for officers to rely on their own experience in laying a gun diminished as their corporate stock of it decreased. Captain R. W. Adye's *The Pocket Gunner* and J. M. Spearman's manual, *The British Gunner*, contained the briefest of range tables in their 1827 and 1828 editions, whereas F. A. Griffiths' *Artillerist's Manual*, first published in 1839, and J. H. Lefroy's *Handbook for Field Service*, which appeared in 1854, both gave handy tabulations which must have had a proportionately beneficial effect on the gun's accuracy of fire. Furthermore the extent of this knowledge was continually being increased: in Griffiths' 1839 edition the capability of a 12-pounder howitzer was only given at 3¾° of elevation but in the 1840 edition it was established that at 5½° it ranged 200 yards further. The 1839 edition recorded only scant figures for the 9-pounder, up to an elevation of 3° or 1,200 yards: in 1847 it was established that with an elevation of 4° it would range 1,400 yards. No tables at all were attempted for the performance of shrapnel in 1839, but they became a feature of subsequent editions.

The combination of experience and observation came also to the aid of gun construction. Artillerymen were faced with a recurrent problem: 'Guns that are made of a certain length have the advantage of throwing shot further, more accurately and with greater velocity . . ., on account of the shot being acted on for a longer time in the same direction while passing through the bore.'[116] This was of course true only up to the point where the whole charge had been burnt up, after which no additional range would be gained. Furthermore, even below this point, doubling the barrel length did not increase the range proportionately but only added about a seventh. Naturally the longer and heavier the gun, the more difficult it was to handle. Thus, although French trials suggested that a gun twenty-seven calibres long gave the maximum range, twenty-two was the norm even for siege guns and eighteen or fourteen for field guns.

It was Gribeauval who had the courage to recognise that the manoeuvrability thus gained justified the loss of range. To enable the lighter equipment to withstand the concussion, he reduced the normal service charge from a half to a third of the weight of the projectile, and in Britain the horse artillery, with its light 6-pounder gun, used only a quarter of the weight.[117] Ballistic justification for these reductions was provided by Sir Alexander Dickson, who calculated that such a charge gave a velocity of 1,340 feet per second. The added initial velocity provided by a high charge – say over 1,600 feet per second – was said to have been reduced to comparable levels after about two or four hundred feet, and by the same token such additional velocity as a gun of twenty-two calibres might impart diminished to that of a fifteen and a half calibre gun within a similar distance. Notable contributions to velocity came in fact from the density of the shot, since the weight lessened the retarding effect of what was then called 'atmospheric pressure', and from a reduction in windage, a third to a quarter of velocity being lost by windage of a twentieth of the diameter of the bore. The fact that the loss of velocity was proportionate to the size of charge and calibre meant that heavier guns lost less to the same windage than did the smaller calibres. Therefore, too, the former were more accurate. A contributory factor here was that, when the gun was loaded, the ball rested on the bottom of the bore, rather than in the centre, and the expended force of the powder tended to pass over the top and not against the centre of the ball. An eccentric ball, to ensure that the axis of rotation and the line of flight could coincide, was impracticable. Therefore, the less the relative windage the greater the accuracy.[118] Dickson reduced the windage in British field guns from about .25 inch to .1 inch,[119] thus permitting a smaller charge and contributing to accuracy as well as velocity, although not apparently to range.

The consistency in range tables and other experiments on patterns of fire suggest that only the best gunpowder was being used for charges and that a uniform standard was thus procured. In fact, the quality of composition and the length of time in store added another imponderable to the gunner's art. Contract gunpowder weighed fifty-five pounds per cubic foot, as against the fifty-seven of that made at the government establishment at Waltham Abbey, and this produced a variation, even when fired in two ounce charges at 240 or 250 yards, of ten or twenty feet.[120] 'You shall take a gun and fire it in pairs of rounds', Dansey wrote, 'each two rounds with the same charge and elevations, you shall then take the mean of each two rounds, and the difference between the longer and the shorter of these two rounds, will be found generally to exceed 1/20 of that mean, frequently to exceed 1/10, occasionally to be as much as 1/4, and sometimes to exceed 1/4 of it.'[121]

To fire the charge, trials since 1816 had been conducted with a view to adapting the percussion system to artillery,[122] and in 1846 two types of percussion tube were approved for use in the army. As late as Chobham camp the traditional portfire (simply prepared composition wrapped in paper, all contained in a tube to enable its insertion into the vent) was still in use. However, by then its days were numbered. The simplest equipment was a friction tube, whereby the pull of a lanyard jerked a piece of metal through detonating composition which in turn sent a spark to the charge. Such a tube was perfected by Mr Tozer of the Royal Laboratory and at trials in 1852 it only missed fire on three out of 400 occasions. The following year it was approved for the artillery.[123] Its essence has remained unchanged to the present day.

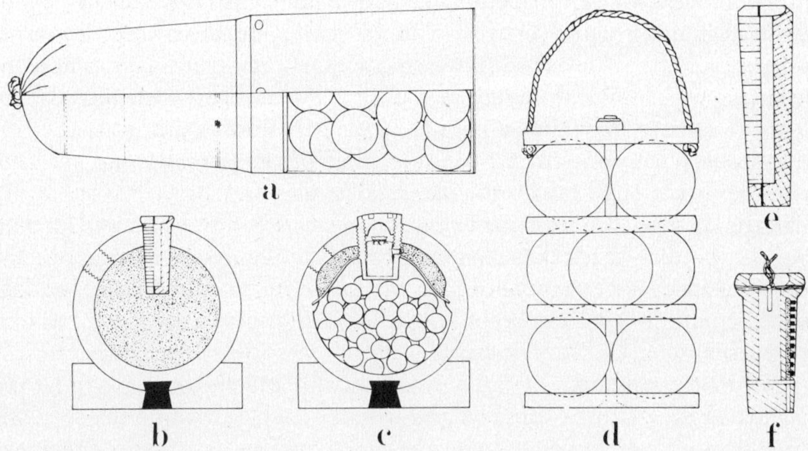


Fig. 18 Types of artillery round: a, canister or case with cartridge bag attached; b, common shell; c, shrapnel with a Boxer fuse; d, grape shot; e, common fuse; f, Boxer fuse. From Robert Wilkinson-Latham, *Crimean Uniforms*, II, *British Artillery* (1973)

The principal types of ammunition fired from British field guns were round shot; canister or case; grape shot; common shell; and spherical case, or, to give it its eponymous title, shrapnel. Round shot was the foremost of these because it was unencumbered with the technical difficulties still associated with shrapnel, and its density gave it the maximum possible ranges – put at 1,760 yards for a 9-pounder at 4¾° elevation.[124] In practice it was rarely fired at greater elevations than 3° in order to encourage it to ricochet off the ground. Particularly at a column of troops, the likelihood of hitting was thereby considerably enhanced, although against a line and with a high trajectory it could bounce over the enemy's heads, as happened with Russian fire at the Alma. Gunners were encouraged to fire round shot at point-blank, not only because even on the bounce it remained remarkably true to direction, but also because frequently the distance from the first to the second graze had proved to be twice that of the original range and instances of four and six times were known.[125]

Lefroy assumed a rate of fire with round shot of two a minute against advancing cavalry. Fire would be opened at 1,000 yards – which he thought the maximum practical range for a 6-pounder – and be maintained until the cavalry was within 300 yards. At that distance, round shot could be replaced by case. Case or canister was made up of numbers of balls, packed into a tin or cylinder, that sprayed out on discharge. Since it was designed for use against large targets in a frontal position, its velocity and range were low. Even against massed cavalry, artillery were cautioned against forsaking round shot too soon. However, a rate of three rounds per minute could be achieved with case, particularly at close ranges, where the elevation was minimal, laying less important and the proximity of the enemy a spur to endeavour. At 150 yards case could be double-charged and gunners were instructed to serve their pieces until the last possible moment, for at such ranges the destruction was appalling.[126]

Grape shot derived its name from its appearance, the cluster of iron balls round a spindle, which thus resembled a bunch of grapes. Its range and fire pattern were similar to those of case, but it was not used in field guns as the bullets could damage the brass bores. It was, however, employed in iron ordnance.

With the popularity of light infantry, and the skirmishing, guerrilla warfare of the colonies, interest and inventiveness settled on the shrapnel shell. Shrapnel might prove some answer to long range rifle-fire from men in dispersed positions. The so-called 'common shell', in which a time fuse exploded a hollow round at its target, sending fragments in all directions, was well established. Its use, however, was limited – partly by the primitive nature of its fuse, which had to be cut to the required burning time before being fitted in the shell and which was therefore difficult to handle in the flurry of action. The lack of an accurate fuse was made all the more frustrating by Shrapnel's invention of 1784. He filled the 'common shell' with shot and thus gave to case the range it lacked when fired in the normal way. The proximity of bursting charge and balls in Shrapnel's shell led to premature explosions and it was said that the flash of the charge could encircle the shell and ignite the fuse in the barrel. Of 1,090 spherical case fired at Woolwich in 1819, seventy-four did burst at the muzzle, and 111 were 'dead'.[127] Griffiths, in his 1839 edition, left the pages on shrapnel blank for the owner to note his own observations. In the same year, therefore, Dickson called for a series of firings with shell at Deal, at ranges from 1,200 to 3,000 yards, to establish more about its capacities.[128] Taken in conjunction with the 1819 trials, the manuals were consequently able to offer clearer guidelines as to the employment of shrapnel fire. While remaining cautious, the *Aide-Mémoire* conceded that in defence, where the distances were known, it was a very good projectile.[129] Its use was countenanced at ranges in excess of 650 yards; at the maximum range for a 6-pounder, 1,100 yards, about 10 per cent of its bullets were effective.[130] The 1819 trials also showed a reasonable spread of hits over targets fifty yards apart. The accepted practice was to aim fifty yards short, so that on exploding the splinters and bullets, volley-like, would carry a further 200 or 300 yards.

The evidence accrued in the 1830s and 1840s argued a less defensive role for shrapnel if its technological imperfections could be overcome. At the crossing of the Chenab in the Second Sikh War, an eye-witness wrote, 'Never was it more clearly demonstrated that shrapnel, when directed with precision, is one of the most formidable and effectual inventions of recent time.'[131] But the armies of Europe, while acknowledging the enormous potential of the shell, were reluctant to adopt it with its existing imperfections. It was therefore fitting that Britain, the original progenitor of shrapnel, should throw up its perfecter. Captain Boxer devised a uniform pattern of fuse, in which two channels were marked in odd and even tenths respectively: in action a hole was bored at the correct mark for the time required. At a trial on shrapnel fired with Boxer's fuse in 1852, 3 per cent were dead shells, 3 per cent burst at the muzzle and 94 per cent took effect; with the old fuse these figures were 16 per cent, 6 per cent and 78 per cent.[132] At the same trial, Boxer's diaphragm shell, in which the bursting charge and the bullets were separated, performed most satisfactorily: in addition to the fifty or so bullets, the casing of each shell broke 'into very large fragments which range great distances'.[133] In March 1853, the Master General approved the manufacture of 14,000 Boxer shells and 16,000 fuses,[134] and soon after agreed that the captain's inventions were an opportunity to put the entire ammunition supply on to a uniform footing.[135] In 1854 a report from Shoeburyness urged the general adoption of both shell and fuse. The troops under orders for the Crimea were rapidly instructed in their use, and Boxer himself was put in charge of their manufacture at the Royal Laboratory.

However, although perfecting an existing system, Boxer's inventions should be considered as the beginnings of a whole new era for artillery rather than as a part of the smoothbore equipment. Until 1850 so easily did the artillery outrange small-arms fire that little pressure was exerted for dramatic improvement. The field gun most commonly in use was the 6-pounder, an extremely light piece, which nonetheless ranged 200 yards at point-blank and 1,200 yards at 4° elevation. The 9-pounder, the preferred ordnance for field if not horse artillery batteries, reached 300 yards at point-blank and 1,400 at 4°.[136] The sighting gear was simple – a tangent sight on the base ring aligned to a fore-sight on the muzzle – and the layer thereby set the elevation with an elevating screw placed, like a jack, underneath the cascabel. With the screw in good order, it was possible for the layer to be accurate to within two minutes. However, much depended on the layer's experience, especially in judging distances. Griffiths's instructions for judging distance emphasise the limited demands made on artillery: masses of troops could only be picked out, he reckoned, at 1,700 yards, and cavalry distinguished from infantry at 1,300. Therefore accuracy against advancing troops was reasonable as soon as they could be distinguished, and at lesser distances the lower trajectory increased the potential target area of troops *en masse*. At the Brown Bess's maximum effective range, field artillery was already at point-blank and problems of elevation ceased to impinge on the gun's accuracy of fire. Therefore, assuming a cavalry charge to take four and a half minutes, and a rate of fire of two rounds a minute to be maintained (or three for case), ten reasonably aimed shots could be discharged. Advancing infantry would be under fire for sixteen minutes and in that time thirty-seven rounds per gun could be fired, including eighteen spherical case, seven round shot, nine canister and three canister double-shotted.[137] One calculation reckoned that one musket ball in 200 found its target, whereas for the more damaging artillery round the proportion was one in ten.[138] Even when the results of a modern study are used to mitigate this theoretical performance – the actual rate of fire in battle seems to have been a round a minute and the maximum effective range 800 or 1,000 yards[139] – a perfected system of smoothbore artillery was more than adequate for the tasks set it.

Therefore pressure for more powerful field artillery or for a new system was checked by the limited tactical tasks which the arm faced. It was not considered part of its role to engage the enemy's artillery in counter-battery fire, and thus was the main spur to invention removed. Wellington only used massed artillery in three engagements – at Busaco, Vittoria and Waterloo – otherwise preferring to place it as support for his infantry. Thus in a cyclical process the British artillery's existing limitations ensured it a lowly tactical role, and vice versa. Even experienced senior officers could therefore afford to misunderstand the guns' capabilities and uses, and gunner officers themselves were not encouraged to take a broader view of their arm's potential. The Woolwich course taught that 'The purposes of artillery in the field are to support the operations of troops',[140] and that 'Artillery never acts independently.'[141] Even such a distinguished ornament to his corps as Lefroy echoed this doctrine in refusing to countenance the concentration of guns.

The result was a strangely warped view of artillery's uses. Jervis, himself too a Woolwich product, forbade the guns to open a battle by firing at the areas of intended attack, since he feared this would reveal the plans to the enemy.[142] Such logic was implicitly counter-productive, particularly when it resulted in the conclusion that preliminary artillery fire only prepared the enemy for a cavalry charge and therefore would rob the latter of surprise. By these tokens the guns had virtually no role in battle at all.

In the Peninsula the forfeited firepower of French columns had been replaced by a fearsome array of artillery. The British line, in maintaining the maximum small-arms fire possible, obscured the deficiencies of the Royal Artillery. Artillery batteries were placed on the wings. There they could not disrupt the cohesion of the infantry line and at the same time they could protect the exposed flanks. Thus divided, they split the enemy's attentions, while still able themselves to reply with converging fire from either direction. The oblique angle of fire took the enemy line at its broadest point and exposed it to a concentrated cross-fire.

The corollary of this positioning was to discourage overhead fire in support of an infantry attack. Jervis advanced two arguments against overhead fire: placing the guns behind the line presented a double target to the enemy, and overhead fire intimidated one's own infantry. However, at Ghuzni the bombardment poured into the fort as the infantry advanced proved to be a tremendous support for the assault. The distribution of two howitzers per battery of six guns should have encouraged overhead fire, Waterloo itself having provided a good example of their efficacy. A howitzer had a short barrel of large calibre, and, with less metal than a normal field gun, its charge could be relatively low and its high trajectory was well adapted to grape and shell. At the Alma, the horse artillery had good results with 24-pounder howitzers at 700 or 800 yards,[143] and at 4° elevation these guns could range 1,050 yards.[144]

The Peninsular school manifested a limited tactical appreciation of the use of field artillery. The guns, predominantly 6-pounders, were light, and yet no attempt was made to transform these into a roving force, to operate on internal lines in Wellington's defensive battles, and therefore to mass at the decisive points. Nowhere was the failure to understand the essence of Gribeauval's and Frederick the Great's work more obvious than in the equipment and draught of the field artillery, paradoxically so good in design but in practice so weak.

With the vast store of guns of various calibres available after the war, British artillery was plagued by a lack of standardisation, and its parts were not interchangeable should its equipment be broken or hit by enemy fire. Each gun carriage, although of the same form, was of different dimensions, giving a total of twelve different types in service.[145] A uniform size of wheel throughout the field artillery provided some insurance, and the 6-pounder gun and 24- and 12-pounder howitzers had a common limber.[146] In 1849, the Master General ordered a committee to establish uniformity in field gun equipment,[147] but on a limited budget it was going to prove a long process to implement any improvement.

Nonetheless, by profiting from others' mistakes, Britain had developed a trail that was considered light, solid and fast. In 1792, Congreve had imported Gribeauval and improved on him. In particular he replaced the double-bracket trail with the block trail, which was lighter, put the balance of the gun nearer the muzzle and thus made it easier to traverse. The wheels employed departed from Gribeauval in all being of an equally large size, so that the balance of the vehicles was good, and the height was well judged to accord with the point of a horse's greatest pulling power. The two-wheeled gun and two-wheeled limber on fair going thus had the speed of a four-wheeled vehicle, while on rough ground the simple pin connection gave the flexibility and cornering inherent in each having a separate carriage.[148] In the same interests of manoeuvrability, England used shafts for the hand wheeler (the off-side horse nearest the wheels) rather than the single central pole favoured by other nations. The latter was easier to draw and better adapted to the yoking of untrained horses, but the former enabled the carriage to turn in a smaller space and the gunners to unlimber the gun with greater speed.

Captain Mathias, writing in the *Berlin Military Journal* in 1818, reckoned that 380.6 pounds was the ideal draught for each horse, and that of the European artilleries the British light 6-pounder came nearest with 539 pounds.[149] In 1849, this, the Royal Horse Artillery's equipment, was still top of the European tables, its power of traction exceeding the weight to be drawn by 235 pounds, and its ammunition waggon being potentially faster than that of any other European power. Even the English 9-pounder had a surplus of 168 pounds, better than any other conventional field artillery, and provided an interesting comparison with the Austrian 6-pounder field gun that with only four horses had to cope with a 253 pounds deficit.[150]

In a sense, the Royal Horse Artillery was either over-horsed or under-gunned. A strong case, discussed below, wished the horse artillery to have 9-, not 6-pounders. The heavier gun was only compatible with the detachment system, in which the gunners were mounted on independent saddle horses and the drivers on the near-side draught horses. Nine-pounders therefore ran counter to a demand for the detachment system to be abolished in the Royal Horse Artillery, the shafts replaced by a pole, and the gunners made to ride on the off-side horses. Fed by Sir Charles Napier, Captain Robertson vehemently espoused this cause in the *Naval and Military Gazette*: his purpose was to release horses for the equipment of more field artillery batteries. The case of the Bengal Horse Artillery was called in evidence, and Napier maintained that in bad country a ridden horse drew a gun better, that poles ensured a more even draught, and that the arrangement gave greater flexibility between the duties of gunners and drivers.[151] He did of course have to conclude that if the arrangement was implemented horse artillery would be limited to 6-pounders, or alternatively be reclassified as ordinary field artillery. It was calculated that bearing six drivers would consume the total efforts of four of the team.

Such a diminution of effort was unacceptable to those who looked to horse artillery to restore to cavalry the powers they had been seeking since the Seven Years War. This school noted with satisfaction the fact that the Bombay Horse Artillery, mounted on the detachment system, performed better than the horse artillery of Bengal in the Afghan War and were able to pursue the Sikhs after the battle of Gujerat. The advocates of the Bengal system would certainly have released more horses for a financially blighted field artillery, but they would also have wrecked the efficiency of the horse artillery. Napier viewed such an outcome with equanimity: 'They only want speed when *charging* and who charges with cannon? Steady well sustained marching at 2 *miles an hour* is the "lightning speed" for me.'[152] Little merit existed in the British army's having the lightest and potentially fastest artillery draught in the world if its generals had no intention of using it.

The denigrators of the Royal Horse Artillery in general, and of the detachment system in particular, must be subject to the same criticism. Assuming the field artillery gunners to be on foot and the total number of pieces in a British force to be proportionately low, only the horse artillery had the ability to move rapidly round the battlefield to the points where it might most be needed. Its potential therefore extended far beyond its accepted tactical role of accompanying cavalry. 'The real object of horse artillery', Colonel Chesney told the 1849 commission, was 'rapid movements, getting on the flank, and covering the passage of rivers and other movements when we have not time to get up heavy guns'.[153] Napier's plodding, two-miles-an-hour field artillery could not take advantage of a suddenly exposed flank or column, or protect its own troops in reversed circumstances. Horse artillery was essentially reserve artillery, able to support a sudden breakthrough or to cover a force in retreat.

However, if this was seen to be its principal task, then it should have been a larger force, and armed with heavier guns, specifically 9-pounders. Alternatively, the foot artillery's gunners should have been mounted on the limbers – as was done in manoeuvres at Woolwich and as subsequently became general practice. Because of the insistence on the gunners walking, the distinction between field and horse artillery, and therefore the continuing relevance of the latter, was enhanced. It meant, too, that the bulk of the guns could not be concentrated rapidly for decisive effect.

Those committed to the disbandment of the Royal Horse Artillery cited instances of field artillery keeping pace with cavalry. Clearly in these cases the gunners were mounted on the limbers and waggons, the accompanying ammunition supply was slender, and, most damning of all, 6-pounders, not 9-pounders, were the guns being drawn. Nine-pounders, properly equipped, could not keep up with cavalry, as was shown at Gujerat, and, if the horse artillery was to retain that as its primary role, 6-pounders were essential. A case therefore existed for the weight of guns used by horse and field artillery to be more sharply divided, to enable the former to concentrate on speed and the latter on metal.

But to the opponents of the horse artillery, this argument confirmed its uselessness. If both arms had 9-pounders, the horse artillery derived its additional speed solely from the detachment system. If one had 9- or 12-pounders and the other 6-, the horse artillery retained speed at the expense of firepower. Ballistically, the 6-pounder could not match the 9-pounder. The latter's shrapnel had normally forty-one musket balls to the former's twenty-seven, its case had 126 to eighty-five, its penetration and range were superior, and it boasted 24 per cent more accuracy.[154] Gough, considering the heavier guns of the Indian native states in 1844, called the 6-pounder a 'pop-gun', fit only to combat infantry, and asked for his horse artillery to have 9-pounders.[155]

A partial solution to the quandary was provided by Captain Grant. Following on earlier work by Sir Alexander Dickson, he diminished the weight at the chase, increased it at the breech, and reduced the windage, with the result that he developed a 9-pounder, which, at ten hundredweight, was only four hundredweight heavier than a 6-pounder and almost four lighter than the 9-pounder used at Waterloo.[156] Trials were first held in 1842–3, but not until 1849 did the Royal Horse Artillery have an experimental issue of the gun.[157] In 1853 it or the conventional 9-pounder seems to have been formally adopted by the horse artillery.[158]

Interestingly, the regimental historian has concluded that the Crimean War did not solve the problem of the best armament for the horse artillery. Sir Frederick Stovin in 1855 thought there should be a positive order against the 6-pounder ever being used again in the field.[159] But, although the gun teams were increased to eight horses, the rigours of service took their toll on the animals' fitness, and on steep ground the 9-pounder proved unable to act fast enough with cavalry.[160] Certainly the increased calibre precluded added stocks of ammunition, and the horse artillery troops at the Alma and Balaclava were forced to curb their rates of fire in consequence.

The debate over the relative merits of 6- and 9-pounders was applicable to all field artillery. Partly because of the vast numbers of 6-pounders in store, and partly because they required fewer horses, the 6-pounder gun and the 12-pounder howitzer were the normal field pieces of the Royal Artillery in the 1840s. However, in a memorandum of 1832, Dickson had proposed a field train for war of five 9-pounders and one 24-pounder howitzer per battery.[161] In the Peninsula, British batteries of six 6-pounders had been worsted by French eight-gun batteries equipped with 8-pounders. Consequently even Wellington was convinced of the

9-pounder's merits,[162] and Gardiner, an ardent 9-pounder man, was an advocate for eight-gun batteries as well. However, not until the arrival of Hardinge and then Raglan at the Ordnance did the Royal Artillery actually have the opportunity to practise with the arm destined for their use in war.

Weight of metal had been impressed on Hardinge not only by the Sikhs' artillery superiority but also by their practice of improving their positions with field fortifications. At Mudki on 18 December 1845 even direct fire by light guns had proved inadequate against an entrenched camp. By the time of Sobraon, two months later, the British had bored out their 9-pounders and enlarged them to 12-pounders.[163] The Sikh Wars showed, it was argued, that large calibres and adequate ammunition supplies were more important than high speeds. Consequently some felt that not even 9-, but 12-pounders should be the normal field guns. Straith tabulated the relative accuracy of a 6-pounder as 12.476 (that is the square of the diameter of shot), of a 9-pounder as 16.483 and of a 12-pounder as 19.642.[164] At a range of 160 yards a 12-pounder shot went through a twenty-foot bank of well-rammed snow and rolled a further eighty-one feet, whereas a 9-pounder stopped after twenty-five feet and a 6-pounder failed even to penetrate.[165]

In France Louis Napoleon advocated that all field artillery should be equipped solely with 12-pounder shell guns. These were weapons capable of firing shells as efficiently as howitzers, and in consequence also called gun-howitzers. The principal advantage in their adoption would be a uniform field gun and standardised equipment. Since the charge was a quarter rather than a third the weight of shot, the gun itself could be lighter. At trials in 1850 a prototype proved more accurate with solid shot than the normal French 8- and 12-pounder guns, and it also possessed comparable powers of penetration. Two years later Chesney drew attention both to Louis Napoleon's shell-guns and to 12-pounder guns. Tactically the Minié had ushered in an age where the howitzer's adaptation to shell-firing would prove vital and where heavier weight of metal would secure the longer ranges at which battles would thenceforth be fought.[166] In April the following year Raglan strove in opposition to his artillery advisers to have the French gun tried against the British 9-pounder. The results, Ross and Colonel Chalmer declared, showed that the 9-pounder ranged further, carried more ammunition and recoiled less. Raglan nonetheless persisted in asking for a 12-pounder battery per division for the Crimea.[167]

Raglan's problem was that as early as the Alma British field artillery had been shown to be grossly undergunned. The Russians employed heavier guns, with the result that they could open fire at distances beyond the capabilities of the British 9-pounders. By nine-thirty on the morning of

Fig. 19 *Royal Artillery: 10 Inch Howitzer & 18 Pounder – Siege Guns.* Lithograph, drawn by G. B. Campion, 1846 (National Army Museum)

Inkerman, three of the six guns in G Battery had been obliged to withdraw from the action, and the remaining 9-pounders were proving 'mere playthings' against the firepower of the Russian 24s. British embarrassment was avoided by summoning 18-pounders from the artillery park, and they proved to be the decisive element in the Russian retreat.[168] The 18-pounder was nominally a gun of position, to support field batteries, and was therefore slow in manoeuvre. But the East India Company used both them and 24-pounders against the Sikhs,[169] and in 1855 Hardinge reported that four 32-pounders had been horsed for the field.[170]

The greater weight of metal employed by European, or European-trained, armies was the product of Napoleon's precepts as to the decisive powers of massed guns, a lesson which Peninsular experience and colonial service had obscured from British ken until the Sikh Wars. It therefore preceded the challenge which the Minié was to pose artillery. The increasing sophistication of artillery found generals schooled in the cavalry or infantry at a loss as to its proper employment. At the same time the failure to promote gunners often left armies bereft of strong professional advice. In Wellington's defensive battles, the guns were required to conserve ammunition and engage the most menacing troops. But on the offensive, as in the Gwalior campaign and in the Sikh Wars, the suppression and neutralisation of the enemy's artillery were essential before an attack could be launched. Colonial, rather than continental experience, thus encouraged a drift to more guns collectively and to heavier guns individually. Hardinge credited the victory at Sobraon to the prior bombardment of the heavy guns,[171] and the success of the final battle of the Second Sikh War, Gujerat, was also attributed to the crushing weight of 100 guns, the largest artillery force until then deployed by a British army. This war convinced Colin Campbell that artillery bombardment was the means to avoiding unnecessary loss of life in the infantry, and so he determined, he wrote in his journal on 21 February 1849, 'to employ this weapon against the enemy to the fullest extent'.[172] Large-scale employment of artillery would make the result more decisive, Colonel Chatterton told the Commons in 1850, the battle shorter and the casualties fewer.[173] Tactically it meant that counter-battery fire became acceptable. If artillery was to be a greater force on the battlefield, defeat of the enemy's guns was a necessary preliminary both for the safety of one's infantry and the attack of the enemy's. Gujerat was opened by both sides' guns engaging each other. One writer concluded that, to ensure victory against massed artillery, its opponents should have a third as many guns again,[174] with the result that the recommended proportions of guns per 1,000 men grew from two or three to six.[175]

The *United Service Journal* in 1831 had seen that, to be decisive, the guns' 'fire should be concentrated, and they should be brought into action in such masses, as to produce at the distance of 700 or 800 yards a paralysing effect. To neglect massing the artillery, or to make it a mere follower of the movements of other troops, shows an utter ignorance of its service.'[176] The same magazine, sixteen years later, published two articles by General Okounef, a Russian officer. Turning the flanks had become the traditional means to victory, he argued, but it led to an orderly retreat, whereas smashing the centre would produce a decisive rout. Artillery, being the arm most open to technological improvement, should be used *en masse* to create a gap in the opposing line. Eighty to a hundred pieces, ideally 12-pounders, should be formed in a concave arc, with ten paces between each gun, to launch an intensive barrage at the enemy; the infantry should walk forward to hold the ground, and the cavalry follow up the enemy's withdrawal.[177] Given the emphasis on field fortification that such a bombardment would encourage, the consequent impetus to yet heavier field guns, and the technological expertise in Victorian Britain to perfect such pieces, the seeds of the First World War were sown. Presciently the *Naval and Military Gazette* saw a revised artillery producing 'a complete revolution in modern tactics'.[178]

Well, therefore, might Sir Frederick Smith announce that 'future battles would be chiefly decided' by artillery.[179] Hardinge's experiences helped that become nearer professional orthodoxy in the artillery, writers such as Inigo Jones and J.-W. Jervis countenancing the massing of artillery in attack, if not defence. Major Straith summed up the new thinking in the introduction to the 1852 edition of his work: 'In all operations in the field, an efficient artillery equipment, is nearly and mainly the arm that decides ... the struggle.'[180]

### (iv) Heavy artillery

Much of the emphasis on the size of guns and their decisiveness sprang not from field artillery, whose growth by 1854 was still in many respects theoretical, but from the traditionally heavier pieces mounted in siege batteries, garrisons and fortresses. When William Malins wrote in 1848 that 'the next war will be decided by great guns', he was talking of French invasion[181] and when Colonel Lewis, R.E., urged that 'guns should be placed to destroy and cripple, and not to teaze', he had in mind coastal defences.[182]

Since the early years of the century heavy artillery – that is to say 18-pounders and upwards – had been cast in iron, rather than the lighter brass still favoured for field guns. A brass gun could probably stand the shock of 2,000 discharges, and then had to be recast. However, to achieve respectable ranges, greater charges were required and only iron gave sufficient strength and accuracy in manufacture. Iron guns were more prone to bursting and required a new vent after three to six hundred rounds,[183] but they had proved essential for the sustained battering of sieges.

In the Napoleonic Wars, heavy artillery had achieved high ranges by firing small masses with great initial velocity. A Frenchman, Paixhans, established that, as transport problems did not arise in coastal defences and other static works, a large mass with reduced velocity could be employed, and thus shell-fire, particularly effective against ships, could be achieved at ranges beyond those hitherto possible with mortars or howitzers. In Britain General Millar was working on Paixhans's principles from 1824: he produced 10-inch and 8-inch shell guns, the former weighing eighty-five hundredweight and initially only for sea service, and the latter fifty hundredweight but replaced in 1838 by larger guns of sixty-five hundredweight.[184] The heavy 8-inch gun ranged 3,010 yards at 12° elevation with a ten-pound charge, and was considered equal in power to the heavy 32-pounder which had hitherto dominated Britain's defences. However, this was only true in terms of range. The 8-inch gun, when firing the sixty-eight-pound shot of which it was capable, stood in terms of

Fig. 20 56-pounder gun, mounted for coast defence. From the title page to Hector Straith, *Treatise on Fortification and Artillery* (1852) (by permission of the Syndics of Cambridge University Library)

relative accuracy in a proportion of 61.622 to 37.699 over the 32-pounder.[185]

Paixhans's and Millar's most immediate contributions were to sea-service ordnance, and it was primarily to this end that a series of trials was instituted at Deal from 1839.[186] However, although the army was not the prime beneficiary, the guns were of course supplied by the Board of Ordnance and the trials conducted by the Royal Artillery. The results, from the land-service viewpoint, were embodied in a table of heavy iron ordnance, showing the specifications of each gun, their ranges at certain elevations, their recoils, and the primary use of each type of equipment. Published on 22 January 1844, it brought order to an area in which the progressive introduction of heavier pieces had led to an excessive proliferation of types and variations.[187]

The principal guns for coast defence were to be 8-inch shell guns, 32-pounders, and Monk's 56-pounder. The latter was the only piece sent for trial specifically for land service between 1828 and 1848.[188] Approved in 1842,[189] the 56-pounder weighed ninety-seven hundredweight and had an eleven-foot barrel. It was nonetheless relatively easily worked as the weight was concentrated away from the chase and round the seat of the charge, so that its strength was comparatively greater for its weight than its predecessors'. In trials in 1839, it had ranged a third further than the

Fig. 21 *Royal Artillery: 68 Pounder in a Casemate*. Lithograph, drawn by G. B. Campion, 1846 (National Army Museum)

42-pounder or 10-inch gun, and at 35° elevation with a seventeen-pound charge it had propelled a lead shell, weighing sixty-three pounds, 5,600 yards. Indeed in 1852, it was believed to hold the record range for a British gun of 5,720 yards.[190]

Yet heavier guns had been approved for sea service, and in 1852 Hardinge, prompted by the fears of French invasion, set up a committee to reconsider the construction and armament of coastal batteries. In particular he was concerned that the 24-pounder might now be considered too light. He wondered whether the 32-pounder should be the smallest calibre, and whether the 68-pounder and 10-inch shell gun might be adopted for land service. With the energy his concern for French intentions gave him, Hardinge instructed the committee to sit three days a week, and within two weeks a revised list of heavy iron ordnance replaced that of 1844.[191] A number of guns, such as the 8-inch gun of fifty hundredweight, were removed from the list, the 24-pounder was retained for batteries protecting shallow beaches, and introduced for the first time were Millar's eighty-six-hundredweight 10-inch shell guns (already used at sea), the 32-pounders of forty-two and twenty-five hundredweight constructed by Monk on the same principles as his 56-pounder, and the 68-pounder of 112 hundredweight. This last gun, Dundas's, could range in excess of

4,000 yards and had performed so well in the Deal trials that it was suggested for the Channel Island defences as early as 1843.[192] Thanks to Prince Albert's intervention, by November 1852, Hardinge had ensured that a hundred 10-inch guns were to be cast, and the following month Parliament voted £14,000 for new iron ordnance, of which only a quarter was for the navy.[193]

In the realm of siege guns, there was a similar drift to heavier pieces and, it may be said, to producing order out of chaos. The 1819 committee of artillery officers established as a basis for future siege operations a train of 140 pieces, including forty 24-pounders, twenty 12-pounders, fifteen howitzers and the rest mortars, of which the 10-inch was the biggest.[194] Sir Alexander Dickson, in his memorandum of 1832, proposed an artillery force of sixty 24-pounders, twelve 10-inch and eighteen 8-inch howitzers, the same for iron mortars, and twenty-five each of 5½-inch and 4⅖-inch brass mortars, making 170 pieces in all.[195]

All three – mortar, howitzer and gun – were designed to play complementary roles. The mortar was a short-barrelled piece. The powder was concentrated into the cone-shaped chamber (called the gomer) so that the whole force of the powder expanded against the tightly-fitting shell. The elevation was normally constant at 45°, range being obtained by variations in the charge. The mortars therefore gave vertical fire. Seventy five° elevation was best for penetration, but this demanded a degree of accuracy that rendered it barely worth while: at 1,000 yards one shot in fifty or sixty hit a target forty feet by eighty feet. Even at 45°, mortars could not be considered precision weapons. On a similar target, at ranges between 600 and 700 yards, 13-inch, 10-inch and 8-inch mortars only averaged 11 per cent hits.[196] Howitzers were used primarily for ricochet, firing in enfilade at elevations of 10° maximum, in order to dismount the guns on the works of fortresses. In trials in 1821, two-thirds of ricochet fire at 400 yards had proved effective, and only half at 600: 600 yards was therefore considered the maximum range for ricochet.[197] Even so, this sort of fire required considerable skill since the proximity of the walls forced the attackers to use a high angle of fire and there was an even chance that the shot would pass clean over the top, striking harmlessly on the other side. Furthermore the construction of traverses countered enfilade fire, and by the 1850s ricochet was thought to be no more than an aid to direct fire, most of the credit going to howitzers in fact belonging to mortars. Finally, 24-pounder guns provided the direct fire, to breach the walls; it was calculated that 10,600 shot fired with full charges at 500 yards were required to make a breach a hundred feet wide.[198]

To open fire at greater ranges, even with a 24-pounder, was construed as a waste of shot and an admission of timidity. Ground was therefore opened out of the defence's range – at Sebastopol it was at 1,600 yards,[199]

Fig. 22 Vauban's system of fortification under siege: the gunners in the foreground are loading a mortar. From the title page to Hector Straith, *Treatise on Fortification and Artillery* (1852) (by permission of the Syndics of Cambridge University Library)

although 600 was more normal. On the first night, the first parallel (a trench running in line with the fortress's front) would be begun and batteries put in place. Thence progress would be made by sapping forward, digging zig-zag approach trenches, until the second and third parallels were completed. The artillery fire from the batteries in the second parallel would breach the walls. It would also cover the preparation of the third parallel, from where the attack would finally be launched.

It was a slow process, but an inexorable one. The system was liable to minor fluctuations only, since most fortresses in Europe were constructed on the same principles. They were derived from the work of Louis XIV's great engineer, Vauban. Vauban's leading idea was to catch the attack in flanking fire by introducing bastions into the enceinte (or wall). However, he thus exposed his own position to enfilade, and the fall of one bastion normally meant the collapse of the whole work. None of his eighteenth- and nineteenth-century successors and improvers, beginning with Cormontaigne, could rid themselves of this presupposition in favour of flanking fire. Montalembert certainly added to the firepower of the defence by building two or three tiers of fortifications, but at the same time he advocated sharper re-entering angles so that the trace of the fort was star-shaped. Carnot, working under Napoleon, also endeavoured to increase the powers of defence by encouraging sorties and using mortars for vertical fire. Vauban's original bastion became protected by lesser works, ravelins and traverses, with salients in the re-entrants, and was itself made as independent from the rest of the fortress as possible. The flanking fire principle was thus sub-divided, and the number of works which the attack was forced to capture increased. Its final form came in the German system of enclosing ground with strong self-contained works, only sometimes linked by an enceinte. The principles of flanking fire in defence and plodding siegecraft in attack were not challenged, but a ring of forts could thus be placed outwith a great city, such as Paris, Linz or Cologne, and preserve the city from direct fire and prolong the operations of the besiegers. Their strength lay not in their individual merits but in their flexibility and combined support.

These works attracted the attention of British observers, and a party of Royal Engineer officers was attached to the Prussian army in 1843–5, for the purpose of taking part in the construction of those at Coblenz, Ulm and Cologne.[200] Each tower was casemated and had two or three tiers of guns, but only the upper was on ground level, the broad ditch creating a sufficient angle to enable the howitzers on the lower floor to fire indirectly.[201] Fort Alexander, the principal work at Coblenz, was protected with subterranean mines, some optimists thinking that the balance would be restored to the defence by countermining. Wellington saw the German system as the only means to fortify London,[202] but the cost of the thirty forts that would be required was put at at least double that of a continuous enceinte, itself estimated at £4 million.[203]

Such an enormous outlay would have been hard to justify. Forty days was thought to be the maximum that any work built on the above lines could hold out.[204] The writings of C. W. Pasley and Sir John Burgoyne, together with a host of lesser imitators, reduced the besieging process to a series of rules. Burgoyne actually drew up an equation for rating each of the main categories of fort, the answer to be multiplied by 'the term of resistance', which he took as only thirty-six days even in the perfect system.[205] 'There is no operation in war so certain as a modern siege', William Napier wrote in his *History*, 'provided the rules of the art are strictly followed.'[206]

In practice, however, British troops rarely had the time or resources to obey the rules. In India, sieges were undertaken in a somewhat cavalier fashion, Charles Napier having only 9-pounder and 24-pounder howitzers with which to assault Hyderabad and Imam Ghar,[207] and Hardinge bemoaning that, of the seventy-six heavy guns and 52,000 rounds required to subdue the Sikh forts, he had only thirty-six and 10,000.[208] In 1849, even the siege of Multan, the nearest to a European fortress, was initially undertaken with thirty-two guns, including twelve 6-pounders. It was renewed with a total of sixty-seven pieces, but only six were 24-pounders: nonetheless, the fortress fell in twenty-seven days.[209] Sir John Keane's army, marching into Afghanistan in 1839, left its field train at Kandahar, and was then faced by the fortress of Ghuzni. Ghuzni had a rampart thirty-five feet high, flanked by numerous towers, and surrounded by a fausse-braie and an unfordable wet ditch. It fell in forty-eight hours. The simple expedient of a bag of powder tied to the gate had covered numerous sins of omission, and was to do so repeatedly in colonial actions against Maori stockades and Chinese forts. So contemptuous did colonial wars make soldiers of the sappers' rules that even escalading seemed to provide instant success if such an attack were launched unexpectedly from several points. Well might a correspondent to the *United Service Magazine* complain that 'the hobby [of fortifications] has been a *little* over-ridden, that it is something like playing a game of Chess with oneself – rather poor profit for the trouble'.[210]

Forts were easily isolated, and if captured could be used by the enemy. If they held out, valuable men had to be committed to their defence and to their relief. They thus robbed a commander of strategic flexibility. Sir George Murray was not alone in his doubts as to the value of forts for home and colonial defence.[211] In 1849 Colonel Thompson, with the support of Sir George de Lacy Evans, told the Commons 'of the superannuated notions on the effect of fortifications, which the experience of modern wars had entirely exploded'.[212] 'He who would occupy forts would be a fool!' coined Sir Charles Napier.[213]

If fortifications were to regain tactical or strategic dominance, they had to have more positive military features than those of delaying the enemy. The initiative had to be restored to the defence. Both Carnot and Montalembert had attempted to give superiority in firepower to the besieged, but in the former's case vertical fire did not prove sufficiently damaging, and in the latter's the tiers of exposed masonry had been easily crumbled by round shot. Straith, who saw the need for a radical departure in the construction of fortresses, recognised that masonry was useless to resist shot, since, even if thick enough, it still sent off dangerous splinters. A low, wide parapet of earth, twenty-five or thirty feet wide at the base, should therefore supersede it.[214] Loosely-packed earth would either absorb the shot or be thrown up only to fall in roughly the same place. The success of the lines of Torres Vedras, and even of the field fortifications at Borodino, reinforced suggestions in favour of more flexible but paradoxically stronger systems of defence.

Straith also mentioned favourably the work of Bordwine, his predecessor as professor of fortification and artillery at the East India

Company's school at Addiscombe. Bordwine condemned angular forts as they were liable to enfilade, their most exposed points – the salients – could not bring direct fire to bear on the enemy, and when the attack had gained the counterscarp each of the bastions was engaged in an isolated battle without the benefits of co-operation. He therefore advocated a circular fort: the enemy would be forced to attack on a much wider scale in order to take the breach, as the defence would be able to pour in continuous direct fire.[215] In 1831 three Royal Engineer officers, including Burgoyne, reported to the Master General that Bordwine's system offered no escape; the lack of salients and flanking fire left the attack's batteries undivided and capable of equally great direct fire, while the fort's own capacity for fire to the front was not, they claimed, greatly enhanced.[216]

It was a civilian architect, James Fergusson, who attempted to draw together the work of Montalembert, Bordwine and others. In a book published in 1849, he noted that sieges were more truly artillery affairs than tests in the engineering skills of sapping, mining and construction. The object for the defence should therefore be to bring to bear the maximum direct fire possible. Fortresses should consequently be circular, and have four ramparts, each fifteen feet above the other and sloping towards the exterior. They should be built of earth. A total circumference of 2,000 yards would thus cost only £62,500, and the money saved could be expended on the 1,000 guns which would be mounted therein. The problems posed for the attackers could be judged by the fact that at the siege of Antwerp the French had only put 144 guns in battery and the largest train yet proposed was 200 guns, which would need 6,000 carriages and 36,000 horses. At any point the defence's guns would be double those of their besiegers. The whole work would be surrounded by a deep and broad wet ditch in order to prevent mining. If a dry ditch had to be resorted to, it should be sixty feet deep. Fergusson reckoned that a string of his forts along the English coast would force the French invaders to divide their forces. Furthermore, since the forts could be manned by auxiliaries, they would forestall any further expenditure on men or ships. The total cost of such a plan he put at £2 million, as opposed to £7–10 million estimated for Colonel Lewis's proposal for a ring of conventional forts between Chatham and Portsmouth, and Canterbury and Chichester. Not only could the principles of Fergusson's system be applied to all types of defences – from entrenched camps upwards – but also the materials could be reused or the structure varied and improved in accordance with the progress of gunnery.[217]

Fergusson exhibited a model of his plan at the Great Exhibition and Lieutenant-Colonel W. H. Adams, professor of fortification at Sandhurst, devoted a whole lecture to the praise of it at the United Service Institution.[218] However, it was the publication in 1852 of *The Perils of Ports mouth*, in which he applied his system to a specific case, that brought Fergusson notoriety. In 1853, Fergusson himself outlined his proposals to the United Service Institution, and four successive evenings were devoted to critiques from distinguished Royal Engineers, including C. W. Pasley, J. Jebb and George Wrottesley, in which they attempted to demonstrate how they would ovecome Fergusson's forts.[219] The balance of opinion following these meetings swung in Fergusson's favour, but only, it was claimed, because the line officers naturally took his side against the Royal Engineers. The debate represented professional vested interests opposed by infantrymen sick of the self-important technicalities of a doomed art.

The essence of the Royal Engineers' case rested on a practicality – insufficient trained gunners would be found to man the defence.[220] Consequently, the fortress was assumed to be smaller than Fergusson intended, and Pasley opposed the defence's reduced weight of metal with an absurdly large train of 550 guns.[221] The attack was to sink its batteries in the ground, and break up the line of its parallels in order to minimise the effect of the defence's fire. The zig-zag approaches would move forward unimpeded by flanking fire. The target presented by the fort would be massive and any shot the besieger did manage to get in would be bound to take effect, the earth from the higher tiers falling in on those below, until eventually a rampart would be made for the assault.

The whole debate was conducted in terms quite outside the experience of European armies, let alone that of Britain. However, it had both specific and general applications in the siege of Sebastopol. The Royal Engineers and the Royal Sappers and Miners had been relatively less hit by peacetime reductions and following reforms in 1812 had an excellent system of practical instruction. Although deprived of a proper establishment of field equipment, increases to the Sappers and Miners reflected those in the Royal Artillery, as the corps's strength rose from 978 in 1836 to 2,063 in 1850.[222] The Sappers were therefore brought to the Crimea at a peak hitherto unsurpassed. The batteries at Sebastopol consisted of more guns collectively and heavier guns individually than in the Peninsular sieges. In the final attack on Sebastopol in September 1855 the British deployed 183 guns. The principal battering gun of earlier wars, the 24-pounder, was deemed too light after the third bombardment, and in the course of the year the numbers of 32-pounders in position grew from seven to sixty-two and of 8-inch guns from eighteen to thirty-three.[223] The naval report on the siege dismissed even the medium 32-pounder as too light, praising the 68-pounders, ninety-five hundredweight, as 'as perfect guns as can be desired ... Their range is very great, and nearly as true as a rifle.'[224] The total number of heavy artillery rounds fired, again from the English trenches alone, was 252,640.[225]

In sum, the British contribution, without considering that of her allies, far outweighed that of any previous siege in recent European history. And yet the siege, instead of lasting the scheduled forty days maximum, dragged on for almost a year. Certainly this was in part caused by the deficiencies of the allies. Burgoyne's advice was broken, the effort was cumulative, the labour protracted, and, with inadequate numbers of men for the extent of the works, disease set in. The quality of the sapping in consequence declined, the confidence of the Russians increased, and the encirclement of Sebastopol was never effected. Assaults were launched from trenches that provided little cover, had not been taken sufficiently close to the target, and were linked with communication trenches inadequate to allow immediate and substantial support.[226] However, the claim advanced by Burgoyne, who had joined in his corps's attack on Fergusson in 1853, that the Russians' success did not vindicate the claim of earthworks, seems shallow. He argued that much of Sebastopol's strength rested on its quantities of artillery,[227] which, particularly as the Russian guns were dispersed, drew a considerable proportion of the allied gunnery effort. But earthworks were another aspect of Sebastopol's strength, and in the assault of 18 June 1855 the abattis of the Russians were found relatively unharmed despite the continuous fire.[228] Horizontal fire had much less effect than vertical on earth parapets. Mortars therefore became increasingly important. The 13-inch and even the 10-inch mortar had previously been considered too heavy for land service, and yet thirty-four of the former and twenty-six of the latter were eventually mounted in the batteries.[229]

### (v) Rockets and rifles

The large increases in the size and number of guns employed in sieges reflected more than the problems posed by innovative techniques in the defence. All the calculations round which Vauban's fronts were reckoned were based on the range of the musket as the lowest common denominator. The introduction of the Enfield rifle meant that the lowest common denominator ceased to be 200 yards and became instead 800 or 1,000. A minor revolution, of which greater weight could be only a part, was required if artillery was to maintain the hegemony on the battlefield which Napoleon had begun to vouchsafe it.

At the peace in 1815, the 'secret weapon' of Europe was the Congreve rocket. It was not very different from a modern firework, and was fired from a tube to give it direction; its performance at Leipzig and Waterloo had been mixed. However, France, Switzerland, Austria and the United States all paid attention to its development in the ensuing years. In Britain, Henry Wilkinson reflected the ambitions of its inventor when he wrote 'I feel confident that in the event of another war, rockets will assume a new and important feature.'[230] In 1852 Chesney, who had gone out to the Balkans in 1829 to equip the Turks with this marvellous device, saw in it part of the answer to the new infantry firepower.

These ambitions rested primarily on the rocket's use as an infantry or cavalry support weapon, rather than as an alternative to thoroughgoing ordnance. Its lightness and portability, its minimal frontage, and the rapidity with which it could be brought into action, all pointed to a flexibility which increasing heaviness and concentration were taking from field artillery. Above all, it was a weapon for broken and difficult country, for mountains and paths over which conventional artillery could not pass. A pack horse could carry seventeen rockets.[231] Against the Kaffirs, therefore, it received extensive employment and it was the only artillery that Sir George Cathcart took with him into the Orange River Territory in 1852.

Many of the more exaggerated claims for rockets sprang from sublime ignorance of their properties. Congreve knew that their speed increased in flight and to this phenomenon he attributed their powers of penetration. Beyond this, knowledge extended little. Griffiths, in his 1839 edition, left the pages on the rocket's performance blank, the *Aide-Mémoire* expressed itself defeated on the subject for lack of reliable information, and Charles Napier found in India that nobody really knew how to use them.[232] In addition, problems began to emerge. The rocket troop of the Bengal Horse Artillery was disbanded in 1828, as extensive employment in the First Burma War had suggested that rockets were sensitive to rough roads.[233] One view was that they ranged most truly on smooth surfaces, provided they were fired from the level they were to traverse. Therefore their capabilities seemed to be diverging from the tactical role they were expected to fill. Furthermore, owing to near-accidents, it was established that rockets did not like extreme temperatures of hot or cold, nor did they last well in store – the former a damning indictment to a colonial army and the latter discouraging to a penurious administration.

Consequently, in 1842, Colonel Dansey was instructed to conduct trials with rockets of all types. In 1844 he drew up a memoir on the preparation and use of rockets, with the results of his trials providing at least some sort of range table. A 24-pounder rocket with all its fuse in and at an elevation of 47° ranged 3,300 yards, while a 3-pounder had maximum and minimum ranges of 1,800 and 420 yards.[234] However, Dansey's report failed to emphasise sufficiently the inaccuracy of the weapon. A warped stick wrecked its line of flight, and in aiming allowance had to be made for the wind. Boxer achieved greater accuracy in 1852 by establishing more stringent methods of production,[235] and Hale developed a rocket which ranged further but had no stick and was therefore easier to transport, less

Fig. 23 *'The Rocket Troop' 1835 ('O' Battery), Royal Horse Artillery.* Lithograph, drawn by W. Heath, *c.* 1835 (National Army Museum)

fragile and less disturbed by the elements. Initially, however, Hale's rockets were rejected, and, although adopted by the United States in 1846, they were not manufactured in Britain until 1854–5.[236]

The doctrine therefore imparted at Woolwich was that rockets would not perform as accurately as guns, but that in favourable circumstances 'they appear capable of being used with good effect against masses of troops, particularly cavalry'. In these cases, 'the very essence and the spirit of the rocket system consists in the facility in firing a number of rounds in a short time, or even simultaneously, with small means'.[237] But massed troops did not appear in the broken country that some still argued was the rocket's forte. In 1847 the rocket troop of the Royal Horse Artillery was broken up and its equipment spread over the other troops.[238] Although rockets lingered on in the British army they were not to prove the answer to the rifle. Great ranges and portability were not matched by accuracy, reliability or even great killing power. A weapon that could frighten massed cavalry would be of little use if cavalry were no longer on the battlefield or no longer massed.

The challenge the Minié and Enfield rifles posed for conventional artillery was as much one of accuracy as of range. At a trial in February 1854, two 6-pounders fired ten shots in 1 min. 20 sec., during which time a Minié fired three rounds and was reloaded. The target was a barrel 600 yards away. The Minié scored a hit and two shots within two feet, while the fourth shot, fired after the time expired, also hit. Not one of the 6-pounders' rounds struck the target and some fell up to 200 yards short.[239] Gun crews were in danger of being so exposed to the fire of skirmishers that the gun would be inoperable before it had even a reasonable target.

Chesney and Sir Howard Douglas were not so pessimistic. Boxer's improvements to spherical case meant that artillery had at least an effective answer to loose formations of skirmishers, and furthermore shrapnel fired from 9- or 12-pounders ranged slightly further than the new rifles. The trend to heavier pieces, and more of them, also suggested that the Minié would not sweep the field unopposed.

However, the heavier the guns, the more circumscribed their use on the field. If the rifling principle could be applied to ordnance, accuracy would not be dependent on weight alone. Rifled barrels would confirm the position of the artillery *vis-à-vis* the other arms in triumphant fashion. The Royal Navy too had an interest in rifled artillery, both for the purposes of long range and accuracy, and for its corollary, a breech-loading mechanism, which on ships would be more convenient than running guns in to load at the muzzle. In 1846 three rifled breech-loading guns of the design of Major Jean Cavalli were ordered to be cast in Sweden for trial by the navy. Also included were three more invented by Baron de Wahrendorff, whose breech-loading system transpired to be an improvement on that of Cavalli. With an eight-pound charge at 15°, the results of the trial were:

| | | |
|---|---|---|
| Service 32-pounder | ranged 3,484 yards, | drift 67 yards |
| Wahrendorff's | ranged 4,247 yards, | drift 131 yards |
| Cavalli's | ranged 4,295 yards, | drift 82 yards |

Nonetheless for universal service, neither Cavalli's nor Wahrendorff's guns could be approved, since the iron had to be specially wrought in Sweden and even then had proved insufficiently strong.[240]

The Ordnance Select Committee, while rejecting Cavalli's work, did at least congratulate him on a successful design: the principles, they felt, were right but English cast iron was not sufficiently strong to allow an adequate breech-loading mechanism to be developed. Encouragement therefore went to the successful production of a rifling system ahead of one for breech-loading. However, a simple application of the Minié principle would not be satisfactory, as the soft lead required to expand into the grooves would not suit the duties of either shell or round shot. Early in 1852, a 9-pounder with four grooves, half an inch deep and half an inch broad, was made, and Lord Clarence Paget devised a conical shell with lead on the base ring only. However trials with 3-, as well as 9-, pounders do not seem to have proved satisfactory. An 8-inch rifled mortar was ordered to be cast and, by 29 March 1853, the Master General could report that experiments on elongated projectiles and rifled cannon were being conducted on an 'extensive scale'.[241]

Foremost among these was the rifling system of Charles Lancaster, already applied to small arms, and which he patented for ordnance in July 1850. In the 68-pounder of ninety-five hundredweight, this took the shape of an elliptical oval bore, the longer axis being half an inch greater than the shorter. The bore had a twist of a quarter of a turn in its length, the twist increasing towards the muzzle. The first trial, with an 8-inch gun, was held in August 1851, and of seven shells, six broke and one ranged 800 or 900 yards.[242] The Master General was unimpressed. Trials in June 1852 on a 6-pounder, rifled on Lancaster's principle, failed to produce results any better than the smoothbore version.[243]

However, perhaps influenced by the intelligence that Belgium was progressing faster with rifling heavy rather than light artillery,[244] Hardinge agreed to the Admiralty's request that trials be renewed with 8-inch shells in a 68-pounder.[245] The shells on this occasion were made of wrought iron and coated with lead, with the result that they could withstand the pressure in the barrel. The practice in trials at the end of 1852 was superb. At 17° elevation and 5,600 yards, the lateral drift was a mere five yards. Admittedly at 15° and 4,400 yards, the drift was fifty yards, but very often it was down to single figures[246] – a degree of accuracy unmatched by any other gun of the period. A 68-pounder, two 8-inch guns of sixty-five hundredweight, two 32-pounders of fifty-eight hundredweight, and one of forty-two hundredweight, were ordered for trial in 1854.[247]

However, before experiments could progress further, both 68-pounders and 8-inch guns bored on Lancaster's principles were mounted in the attack on Sebastopol. Initial reactions were mixed. Their range, the sound of the shell – likened to an express train – and their killing power were praised. But Burgoyne was sceptical, and, although he attributed it to inexperience in their use, Raglan reported that the Lancasters were proving 'very wild and ungovernable in their flight'.[248] Views as to the best method of employing them varied as much as their performance: round shot was found to be more accurate at short ranges and shells at longer; some argued for heavy charges and heavy shot, others for light charges and light shell.[249] The ammunition supply was limited, and, although a special Lancaster shell factory was set up at the Royal Arsenal in November,[250] two guns firing a round every eight minutes were not enough to confirm any diagnosis in the interim.

However, trials at Shoeburyness in October 1854 showed a uniform tendency in both the 68-pounder and the 8-inch to drift to the right. The extent of the error varied, and was as high as sixty-five yards at 2,800 yards, but only twenty-five yards at 4,923 yards. The general conclusion confirmed the growing opinion in the Crimea that 'the Ranges with the same elevation and charges vary considerably, and consequently no medium of five, or any other number of rounds can be depended on'.[251] One gun burst at Shoeburyness and a further three did so at the Crimea. Major Mordecai of the United States Army concluded that the wrought iron shell was now too powerful in relation to the barrel,[252] and this view was confirmed by the use of ordinary 8-inch spherical shells in the Lancasters at Sebastopol.

Naval interest in the Lancaster was maintained for a couple more years, but for land service the principle was eclipsed in 1855 by the work of Armstrong and Whitworth. Armstrong used wrought iron for his barrels, so that they could take the strain of an elongated projectile. The result was a field gun that was not only rifled but also breech-loading. The problems of applying systems effective in a rifle to ordnance were proportionately greater, and the time lag in technological development can be judged as finely as the gap between the rifle-dominated battle of Königgrätz and the field guns of Sedan. The contribution of the Lancasters to the Crimean War was, particularly by the standards of rifled artillery in 1870, negligible. But their appearance at all is evidence both of a British desire to experiment and innovate, and of the artillery's response to the challenge of the infantry. It does not seem illogical to relate the heavy guns of the attack and the resilient earthworks of the defence to the events of a war then still sixty years in the future. Furthermore, this projection is substantiated by Boxer's refinement of the shrapnel shell. Herein was a weapon which by its dominance of open ground would force infantry to forsake even a skirmishing formation, and to dig deep networks of trenches in the search for self-preservation.

An arm that had been consistently mishandled and misunderstood by Wellington had gained a more coherent tactical role, together with its concomitants, greater mobility and an increased weight of metal. And the factors that had propelled the British artillery into the nineteenth century were only to a limited extent the precepts of Napoleon and the challenge of the Minié rifle. As important as the pressures from Europe were the advances in British technology and the acquisition of a range at Shoeburyness where they could be properly tested and improved. Furthermore these refinements were determined less by scientists and manufacturers than by the demand and experimentation of Royal Artillery officers themselves. Boxer was responding to a known need, the perfection of the shrapnel shell. Even the development of rifled breech-loading ordnance, in which the great manufacturers, Lancaster, Armstrong and Whitworth, took the lead, was a response to a known desideratum in the regiment.

Finally the pressure which clinched many arguments, and which produced in Hardinge a Master General of the Ordnance committed to improved artillery, derived from colonial experience, from the Sikh Wars, in which paradoxically no unit of the Royal Artillery had been directly engaged.

# Conclusion

The expeditionary force that sailed from Britain in 1854 was neither technologically nor tactically inferior to its opponents. Indeed, together with the French, it was eventually able to defeat an army whose prowess had coloured international relations since 1815. Furthermore the debates and developments described in the preceding chapters give the lie to this being the Peninsular army, slightly refurbished but nonetheless tarnished by age.

There were three major reasons why the years 1815 to 1854 were not barren for British military thought. The first was that the campaigns of the Peninsula and Waterloo were themselves revolutionary: Britain's effort against Napoleon was not confined to commercial and naval pressure, but embraced a major military effort on the European mainland. The implications for, and effects on, the army were bound to require analysis. Secondly, the victory of 1815 confirmed Britain's hegemony overseas: over the next fifty years the empire increased at the rate of 100,000 square miles a year.[1] Continentalism had therefore to be accompanied by colonialism. Finally, particularly towards the end of the period, technological innovations began to transform the ground rules of tactics in a fashion more dramatic than at any stage since Marlborough.

The debates that ensued were fuelled, in particular, by two classes of officer. First were those of the Royal Artillery and Royal Engineers, who had themselves received a technical education, and who were therefore poised to welcome the advent of the rifle and the breech-loader. Second were the Peninsular veterans, now on half-pay, who passed their retirement in correspondence with the military press. The consequent burgeoning of military literature under such influences was a phenomenon that also occurred elsewhere, and especially, as other writers have shown, in France and the United States.[2]

In practice, the British army's problems between 1815 and 1854 were as much those of innovation as of stagnation. British military thought was more active and more sustained than ever before in the history of the army. The troops that sailed in 1854 were not short of ideas, but they had not ranked them: they had no settled doctrine. Colonial obligations made it extraordinarily difficult for commanders to anticipate or to plan. As one of Britain's leading military thinkers, G. F. R. Henderson, put it in 1900: 'It is useless to anticipate in what quarter of the globe our troops may be next employed as to guess at the tactics, the armament and even the colour . . . of our next enemy. Each new expedition demands special equipment, special methods of supply and special tactical devices, and sometimes special armament.'[3]

What the army had never anticipated between 1815 and 1854 was the possibility of conducting offensive operations against a continental opponent. If they were to fight in Europe, it would be to defend the United Kingdom against French invasion. But in the empire they had encountered a wide variety of warfare: they had met the Sikhs on terms and in battles that resembled those against Napoleon; they had fought in the mountains of Afghanistan, and in the bush of Kaffraria and New Zealand; they had campaigned in the cold of Canada and the torrid heat of Baluchistan. Such experiences begat ideas. Although much of military theory was Eurocentric, it was refined by colonial practice. Nor were the consequent conclusions clear-cut. Some types of operation encouraged loose, skirmishing formations, others the serried line. Evidence from one set of European examples and from certain colonial wars might support one position; other European examples and other colonial wars might lend weight to the opposite. The difficulty of drawing conclusions about the future pattern of British military operations was immense.

The essence was flexibility. These contradictory pressures and influences at least familiarised British officers with the value of mounted infantry, of massed heavy artillery, and of looser infantry formations. The innovations that appeared between 1815 and 1854 were to be swamped by the intensity and rapidity of change in the second half of the century. But we should not therefore blind ourselves to the developments of the 'long peace'.

# Notes

Full titles and bibliographical details of all works are given in the bibliography.

### 1 The theory and the conduct of war

1 W.O. 99/19.
2 Bodleian MS Eng. Lett. c 252, f 174, W. Napier to Jervis, March 1852.
3 Hardinge papers, Yates to Hardinge, 16 October 1852.
4 *Edinburgh Review*, XXXV, 1821, 377.
5 Bruce (ed.), *Life of General Sir William Napier*, II, pp. 227–8.
6 N.L.S. Adv MSS 46.9.15, p. 183, Murray to Henry Drummond, 17 January 1845.
7 W. F. P. Napier, *C. J. Napier*, I, p. 450.
8 *London and Westminster Review*, VI and XXVIII, 1838, 379, 401.
9 Jomini, *Art of War*, pp. 34–5.
10 MacDougall, *Theory of War*, p. 156.
11 Hamley, *Operations of War*, pp. 422, 324, 37.
12 Burgoyne papers, EC 72. The paper seems to have been written soon after 1813, the most recent date referred to in it.
13 Sir George Cathcart papers, 'Essay on Strategy, May 1831, Halifax, N.S.'.
14 *Aide-Mémoire to the Military Sciences*, I, p. 1.
15 Luvaas, *Education of an Army*, p. 19.
16 *U.S.M.*, 1852, III, 557–8.
17 Wellington papers, Liverpool to Wellington, 10 September 1840; Gurwood to Wellington, 22 September 1840.
18 *U.S.G.*, 20 June 1840, 4.
19 John Mitchell, *Thoughts on Tactics*, p. 121.
20 Ibid., p. 12; also p. 122.
21 Ibid., p. 7.
22 *U.S.G.*, 14 May 1842, 4.
23 Ibid., 3 June 1854, 4.
24 *U.S.J.*, 1838, II, 51.
25 John Mitchell, *Thoughts on Tactics*, pp. 4, 5, 147.
26 Ibid., p. 120; it is instructive to compare this with Clausewitz's chapter on 'the nature of battle today', *On War*, ed. and trans. by Michael Howard and Peter Paret, Princeton, 1976, book 4, chapter 2.
27 Ibid., pp. x–xi.
28 Ibid., p. 147.

29 Ibid., p. 9.
30 Ibid., p. 161.
31 Ibid., p. 79.
32 Ibid., p. 122.
33 *Edinburgh Review*, XXXV, 1821, 378.
34 See Strachan and Barthorp, 'The Battle of Meanee', 127–34.
35 Lehmann, *Remember You Are an Englishman*, p. 218.
36 Griffith, 'Military Thought in the French Army', 20, 42, 69–70, 128, 175–7, 256.
37 *Aide-Mémoire to the Military Sciences*, I, p. 2.
38 Ibid., p. 10.
39 Hew Strachan, 'Soldiers, Strategy and Sebastopol', 303–25.

### 2 Infantry: the rulers of the field

1 *U.S.J.*, 1829, II, 415.
2 Archives Historiques de Guerre, MR 1430/8, p. 15, report by Col. de Lostende, 12 August 1835.
3 *N. & M.G.*, 8 October 1836, 646; 15 October 1836, 663; 29 October 1836, 703.
4 Seaton papers, Hardinge to Seaton, 10 June [1853].
5 N.L.S. MSS 1849, ff 9–27, memo by Brown, 15 February 1854.
6 *Memoranda by Lieut.-General Lord Frederick Fitzclarence, for the Use of Young Officers Assembled in Poona*, p. 2.
7 *U.S.M.*, 1847, II, p. 329.
8 Journal of Lt. G. A. Bace, *Journal of the Society for Army Historical Research*, XXX, 1952, 57.
9 W. J. B. Mountsteven, *Observations on the Infantry Manoeuvres of the British Army*, p. 18.
10 N.A.M. 7709–6–11, pp. 187–90, Vivian to Blakeney, 14 March 1833.
11 Anderson, *Recollections of a Peninsular Veteran*, pp. 140–1.
12 *Journal of the Society for Army Historical Research*, XXXII, 1954, 41.
13 Archives Historiques de Guerre, MR 1430/20, pp. 22–3.
14 N.L.S. MSS 1848, f 55, Cochrane to Brown, 23 July 1852.
15 C. W. C. Oman, *Wellington's Army*, p. 84.
16 *N. & M.G.*, 20 September 1834, 695.
17 Fitzclarence, *Suggestions for Brigade and Light Infantry Movements*, pp. 17–19.
18 East Yorkshire C.R.O. DDCH 80, p. 259, diary of Sir Charles Chichester, 16 November 1843.
19 [Ryder], *Four Years' Service in India*, p. 117.
20 Atkinson, *South Wales Borderers*, p. 302.
21 *Royal Military Magazine*, 1852, I, 83.
22 N.L.S. MSS 2849, f 68, Cochrane to Brown, 23 August 1851.
23 Troubridge, *Principles of the Minor Operations of War*, p. 40.
24 W.O. 135/3, memo by Smith, 24 February 1844, para. 86 (copy).
25 Le Couteur, *The Rifle*, p. 54.
26 W.O. 80/3, Murray to Peel, 25 August 1845.
27 B.M. Add MSS 40461, ff 197–9, Wellington to Peel, 28 August 1845.
28 *U.S.J.*, 1831, II, 205.

29 *N. & M.G.*, 9 April 1842, 230.
30 Jervis, *Manual of Field Operations*, p. 62.
31 Lopez Suasso Diaz de Fonseca, *Theory of Infantry Movements*, I, p. xxxv.
32 S.R.O. GD 225/box 42/packet, A. Leith Hay to his mother, 27 October 1854.
33 *Minutes of Proceedings of the Royal Artillery Institution*, I, 1857, 24.
34 Lendy, *Principles of War*, pp. 16–17.
35 Mitchell in *U.S.J.* 1831, II, 1, and 1834, II, 309; W.O. 80/3, Murray to Peel, 25 August 1845.
36 *British Army Despatch*, 1 July 1853, 423.
37 *U.S.G.*, 25 March 1837; *N. & M.G.*, 29 November 1851, 762.
38 *Memoranda by Lieut.-General Lord Frederick Fitzclarence, for the Use of Young Officers Assembled in Poona*, p. 3.
39 *Aide-Mémoire to the Military Sciences*, III, p. 633.
40 N.L.S. MSS 2845, ff 13–15, A. Hay to Col. Felix, 14 January 1846.
41 Knollys (ed.), *Life of General Sir Hope Grant*, I, pp. 96–8.
42 Ibid.
43 B.M. Add MSS 49107, f 20, C. Napier to Robertson, 29 October [1838].
44 *The Times*, 27 December 1886, quoted in Rait, *Life of Gough*, II, pp. 57–8, 276.
45 *U.S.J.*, 1838, II, 187.
46 Ibid., 1839, III, 531; 1840, I, 262.
47 *N. & M.G.*, 4 January 1840, 5.
48 Kinloch, *Proposal for the Defence of the Country*, p. 46.
49 *U.S.J.*, 1839, III, 249.
50 Ibid., 1840, I, 108.
51 *N. & M.G.*, 18 February 1837, 101.
52 *U.S.G.*, 26 March 1853, 7.
53 King, *Campaigning in Kaffirland*, p. 90.
54 W.O. 3/110, p. 409.
55 Hardinge papers, E. A. Angelo to Hardinge, 23 April 1853.
56 *N. & M.G.*, 5 October 1839, 639.
57 *Journal of the Society for Army Historical Research*, XVIII, 1939, 195.
58 Cowper (ed.), *The King's Own*, II, p. 63.
59 W.O. 44/539.
60 W.O. 44/677, ff 57–115; 3/291, p. 233; 3/299, pp. 198, 421; 3/300, pp. 28, 63, 127, 128.
61 *N. & M.G.*, 6 September 1853, 504.
62 William Barwick Hodge, *Journal of the Statistical Society of London*, XIX, 1856, 255.
63 East Yorkshire C.R.O., DDCH 95, p. 26, C. R. Chichester's diary, 18 September 1846.
64 Wellington papers, Wellington to Somerset, 9 November 1835.
65 *Aide-Mémoire to the Military Sciences*, II, pp. 425–7. Windage is the gap between the ball and the side of the barrel, or, put in different terms, the difference between the diameter of the ball and that of the bore.
66 W.O. 135/3, Smith to Kempt, 9 August 1842.
67 Bruce (ed.), *Life of General Sir William Napier*, II, p. 379.
68 *U.S.J.*, 1830, II, p. 598.
69 Hughes, *Firepower*, pp. 98–9, 103, 122, 124, 127.
70 Copies of the report are in P.R.O. 30/46/2/4 and W.O. 1/477, p. 451.

71 Second Earl Cathcart papers, J 26, Brown to Cathcart, 8 January 1852.
72 Wrottesley, *Life of Burgoyne*, I, pp. 429–30.
73 *N. & M.G.*, 12 October 1850, 652–3.
74 *U.S.J.*, 1834, II, 323.
75 W.O. 44/701.
76 W.O. 44/133.
77 P.P. 1849, IX, 291.
78 *N. & M.G.*, 29 March 1834, 202–3.
79 *U.S.G.*, 6 September 1834.
80 W.O. 44/304.
81 W.O. 44/677, pp. 1002–38.
82 W.O. 44/520.
83 W.O. 44/677, f 197; Hansard, 3rd series, XLII, c 626, gives 200,000 in store in 1838, as against 900,000.
84 W.O. 44/133.
85 W.O. 44/677, ff 211, 567–70.
86 B.M. Add MSS 40474, ff 80–1, Hardinge to Peel, 28 June 1843; ibid., 40460, f 264.
87 W.O. 80/3, Murray to Peel, 27 June 1845, 6 July 1845.
88 Wellington papers, Anglesey to Wellington, 21 November 1846.
89 P.P. 1849, IX, pp. 263–4.
90 Wrottesley, *Life of Burgoyne*, I, p. 493.
91 W.O. 3/105, p. 234.
92 W.O. 3/109, p. 244.
93 W.O. 44/677, ff 295, 301, 423; B.M. Add MSS 54519, ff 136, 143, Pennefather to C. Napier, n.d., 8 August 1845; N.L.S. MSS 2843, ff 84–5, R. H. Wynyard to Brown, 3 July 1843.
94 N.A.M. 7709–6–15, pp. 127–8, Vivian to Sir C. Napier, R.N., 20 October 1841.
95 W.O. 3/317, p. 422.
96 Blackmore, *British Military Firearms*, p. 226.
97 P.P. 1849, IX, 289.
98 *U.S.J.*, 1830, II, 598.
99 W.O. 44/132.
100 Sanderson (ed.), *The Arthur Papers*, letter no. 1140; W.O. 3/95, p. 171; W.O. 3/290, pp. 194, 284.
101 East Yorkshire C.R.O. DDCH 81, p. 148, Sir Charles Chichester's diary, 28 August 1844.
102 W.O. 44/677, ff 574–689; W.O. 44/520.
103 W.O. 33/1, p. 554.
104 W.O. 3/283, p. 38; W.O. 3/289, p. 423; W.O. 44/677, f 871.
105 N.L.S. MSS 2846, f 41, Buller to Brown, 20 February 1847.
106 *Journal of Society for Army Historical Research*, XVII, 1938, 112.
107 Archives Historiques de Guerre, MR 2042/30, p. 11; see also p. 13, and MR 2042/38, pp. 13, 42.
108 S.R.O. GD 45/8/86, pp. 11–12, F. Smith to Maule, 16 November 1850.
109 W.O. 3/313, p. 61.
110 Douglas, *A Treatise on Naval Gunnery*, p. 343.
111 *U.S.G.*, 8 December 1849, 5.
112 Douglas, *A Treatise on Naval Gunnery*, pp. 356–7.

113 Hardinge papers, Simpson to Hardinge, 28 October 1852.
114 For a detailed discussion of the evolution and trials of the Minié and related arms, see Roads, *The British Soldier's Firearm 1850–1864*, and the same author's unpublished thesis, 'The History of the Introduction of the Percussion Breech-Loading Rifle into British Military Service 1850–70'. There are also full articles on the evolution of the arm in France in the *Military Review*, I, and II, 1852–3.
115 P.P. 1854, XVIII, 274.
116 Herbert papers, Wellington to Anglesey, 13 May 1851 (copy); Board of Ordnance minute quoted in Blackmore, *British Military Firearms*, p. 229.
117 R.A. E 43/4.
118 W.O. 46/86, p. 142.
119 P.P. 1854–5, IX, part III, *4th Report*, 322.
120 W.O. 3/316, p. 138.
121 P.P. 1860, VII, 389, 455; Wellington papers, Wellington to Russell, 29 January 1852.
122 Hugh Robinson's letters, 15 April and 15 May 1852.
123 Horse Guards circulars, 24 July 1852, 13 January 1853; W.O. 3/114, p. 257; W.O. 3/319, p. 110.
124 W.O. 3/115, pp. 429, 472; W.O. 3/332, p. 314.
125 Curtiss, *The Russian Army Under Nicholas I*, p. 127.
126 Letter of H. Dixon, quoted in Sotheby catalogue, sale of 25 March 1974, lot 292.
127 *Royal Engineers Professional Papers*, new series, IV, 90.
128 *Military Review*, I, 1852, 174.
129 P.P. 1854, XVIII, 498.
130 W.O. 44/701.
131 Hansard, 3rd series, CXXX, c 1290; Mordecai, *Military Commission to Europe, in 1855 and 1856*, p. 211.
132 *Military Review*, I, 1852, 132.
133 N.A.M. 6807/289, Hardinge to Raglan, 8 April 1854.
134 Third Earl Grey papers, Russell to Grey, 16 February 1852; Hansard, 3rd series, CXX, c 184; Herbert papers, Wellington to Anglesey, 7 October 1851 (copy).
135 W.O. 3/317, pp. 137, 422; N.L.S. MSS 1848, f 169, Cochrane to Brown, 21 November 1853; Gordon, *Remarks on National Defence*, appendix, p. 10.
136 P.P. 1854–5, IX, *2nd Report*, 255.
137 Douglas, *A Treatise on Naval Gunnery*, pp. 346–9, 600–6.
138 Hansard, 3rd series, CXX, c 186.
139 *Edinburgh Review*, XCVI, 1852, 523.
140 *U.S.M.*, 1852, I, 98, 579.
141 *Quarterly Review*, XC, 1852, 491.
142 B.M. Add MSS 49105, ff 75–6, C. Napier to Ellenborough, 27 September 1851; C. J. Napier, *Defects, Civil and Military, of the Indian Government*, pp. 303–6.
143 N.A.M. 6210/95, Estcourt to Wetherall, 13 December 1854.
144 King, *Campaigning in Kaffirland*, p. 214.
145 Wrottesley, *Life of Burgoyne*, II, p. 479.
146 W.O. 46/88, p. 6.
147 W.O. 46/89, p. 92.

148 Clark-Kennedy papers, Military Secretary to Major John Clark-Kennedy, September 1855.
149 W.O. 46/89, p. 91.
150 Gordon, *Remarks on National Defence*, appendix.
151 W.O. 3/320, p. 355.
152 W.O. 3/320, p. 421; W.O. 44/701.
153 W.O. 3/321, p. 429; W.O. 33/1, p. 561.
154 W.O. 3/324, pp. 13, 65; 3/325, p. 23.
155 Hardinge papers, Hardinge to Prince Albert, 29 August 1855.
156 Hardinge papers, Hardinge to Palmerston, 15 November 1854.
157 Jervis, *The Rifle Musket*, p. viii.
158 W.O. 33/1, pp. 553, 561–5.
159 Roads, *The British Soldier's Firearm 1850–1864*, p. 69.
160 Hereford C.R.O., G/IV/A/423, Airey papers, Hardinge to Airey, 23 March 1855.
161 Hardinge papers, Hardinge to Prince Albert, 29 August 1855.
162 N.L.S. MSS 2839, f 108, Brown to Gardiner, 26 July 1834.
163 Jervis, *Manual of Field Operations*, p. 73; *U.S.M.*, 1843, III, 447.
164 Major J. P. Sparks in *U.S.M.*, 1846, II, 239, and 1847, II, 291; East Yorkshire C.R.O., DDCH 84, p. 67, diary of Sir Charles Chichester, 23 September 1846; Chichester's results seem to be those published in *N. & M.G.*, 19 December 1846, 814.
165 *Manual and Platoon Exercises*, 1834, pp. 43–50.
166 *Queen's Regulations*, 1844, p. 109.
167 Hansard, 3rd series, LXVII, c 331.
168 N.L.S. MSS 2843, f 64, R. Irton to Brown, 30 April 1843.
169 J. Bentham in *U.S.M.*, 1849, I, 415.
170 East Yorkshire C.R.O., DDCH 79, pp. 68–9, diary of Sir Charles Chichester, 18 February 1842.
171 Ibid., 84, p. 38, 22 August 1846.
172 *U.S.G.*, 15 April 1837, 25 May 1839, 6; 10 August 1839, 5.
173 Ibid., 13 February 1841; *U.S.M.*, 1849, I, 412.
174 *Aide-Mémoire to the Military Sciences*, II, pp. 431–5.
175 Wrottesley, *Life of Burgoyne*, I, pp. 429–31.
176 S.R.O. GD 45/8/86, pp. 53–7, F. Smith to Maule, 16 November 1850.
177 W.O. 3/317, p. 193.
178 Horse Guards circular, 24 July 1852; W.O. 3/317, p. 401; W.O. 3/319, p. 110.
179 N.L.S. MSS 9872, ff 58–9, Pasley's diary, 29 March 1852.
180 R.A. E 2/51, E 44/89, E 2/61, Hardinge to Prince Albert, 20 January, 31 January, 11 February 1853; N.L.S. MSS 1849, f 14, memo by Brown, 15 February 1854.
181 W.O. 3/320, p. 115.
182 W.O. 33/1, pp. 545–7.
183 Clark-Kennedy papers, John Clark-Kennedy to his mother, November 1854.
184 W.O. 3/322, p. 113; W.O. 3/115, p. 431.
185 *U.S.G.*, 2 March 1850, 5; 21 December 1850, 3; 22 March 1851, 3.
186 W.O. 3/321, pp. 115–21.
187 W.O. 3/320, pp. 181, 208.
188 Hardinge papers, Hardinge to Prince Albert, 20 January 1854; Ross-of-Blandensburg, *A History of the Coldstream Guards*, p. 127.

189 W.O. 3/116, pp. 128, 279, 284, 305, 343, 444, 458, 478, 482.
190 W.O. 3/116, p. 45.
191 R.A. Add E/1, no. 169, memo by Wetherall, 22 May 1854.
192 N.L.S. MSS 1848, f 172, Hardinge to Brown, 30 November [?1853].
193 Seaton papers, undated memorandum, possibly that referred to in a letter to Hardinge, 15 September [1853].
194 *Military Review*, II, 1853, 1.
195 R.A. E 2/51, Hardinge to Prince Albert, 20 January 1853.
196 Maitland, *Theoretical Instruction in Ball Practice*, p. 49.
197 Le Couteur, *The Rifle*, pp. 28–9.
198 Bodleian MS Eng. Lett. c 253, f 31, W. Napier to editor, *N. & M.G.*, 1853.
199 Chesney, *Observations on the Past and Present State of Fire-arms*, pp. 286–9.
200 G. Gawler, *Essentials of Good Skirmishing*, 1852, pp. 35–41.
201 Hereford C.R.O. G/IV/A/385, Hardinge to Airey, 23 September 1854.
202 R.A. Add E/1 no. 152a, memorandum by command of Seaton, signed R. Wood, 3 June 1853.
203 Albert Seaton, *The Crimean War*, pp. 89, 101.
204 *Journal of the Society for Army Historical Research*, XVI, 1937, 120.

### 3 Cavalry: the blighted flower

1 *U.S.J.*, 1831, I, 305.
2 Lt. Col. Gurwood (ed.), *Dispatches of Field Marshal the Duke of Wellington*, London, 1838, VII, p. 374.
3 Bodleian MS Eng. Lett. d 240, f 9, Brotherton to W. Napier, 26 February [?1831].
4 F. de Brack, *Advance Posts of Light Cavalry*, p. 174.
5 Von Bismark, *Lectures on the Tactics of Cavalry*, pp. 88, 92.
6 Oman, *Wellington's Army*, p. 101.
7 B.M. Add MSS 54561, f 44, memo by C. Napier, 30 October 1840.
8 Gawler, *Essentials of Good Skirmishing*, 1852, p. 28.
9 Bodleian MS Eng. Lett., d 239, f 38, F. Ponsonby to A. F. Barnard, 1 November [?1830].
10 *British Army Despatch*, 11 April 1851, 232.
11 *U.S.M.*, 1842, I, 209.
12 W.O. 3/90, p. 429; 3/284, p. 120; 3/285, p. 38; 3/290, p. 131; 3/291, p. 417; 3/293, p. 171.
13 Bismark, *Lectures on the Tactics of Cavalry*, pp. 141–2.
14 *U.S.M.*, 1850, III, 14.
15 *U.S.G.*, 1 April 1854, 5.
16 Marquess of Anglesey, *History of the British Cavalry*, I, pp. 164–5.
17 Portland Collection, PwJf 2210, Whittingham to Bentinck, 6 May 1831.
18 Bismark, *Lectures on the Tactics of Cavalry*, p. 143.
19 *N. & M.G.*, 19 November 1836, 749.
20 R.A. Add E/1, no. 155; Verner, *Military Life of Duke of Cambridge*, I, pp. 54–5.
21 N.L.S. MSS 2846, ff 126–7, Cambridge to Brown, 7 November 1847.
22 *Military Review*, II, 1853, 193; Kinloch, *Proposal for the Defence of the Country*, p. 17.
23 *Regulations for the Instruction, Formations and Movements of the Cavalry*, 1833, p. 171.

24 Humbley, *Journal of a Cavalry Officer*, pp. 145–6.
25 W.O. 3/115, p. 184.
26 Third Earl Grey papers, Smith to Grey, 7 November 1851.
27 Cathcart, *Correspondence of Lieut-General the Hon. Sir George Cathcart*, p. 345.
28 Nolan, *Cavalry: its History and Tactics*, pp. 124, 127.
29 Bolitho, *The Galloping Third*, p. 150.
30 Bisset, *Sport and War*, p. 91.
31 *Journal of the Society for Army Historical Research*, XIX, 1949, 194; *Cavalry Journal*, II, 1907, 461.
32 *N. & M.G.*, 6 June 1846, 365.
33 W.O. 44/539; P.P. 1850, 267.
34 W.O. 44/701.
35 Evelyn Wood, *Crimea in 1854 and 1894*, p. 113.
36 Williams, *Historical Records of the Eleventh Hussars*, p. 217.
37 Luard, *History of the Dress of the British Soldier*, pp. 128–9.
38 C. J. Napier, *Defects, Civil and Military, of the Indian Government*, p. 300.
39 Bismark, *Lectures on the Tactics of Cavalry*, pp. 325–6.
40 Bolitho, *Galloping Third*, pp. 142–3.
41 Nolan, *The Training of Cavalry Remount Horses*, p. 35.
42 Henry Angelo, *Instructions for the Sword Exercise*, p. 19.
43 W.O. 44/677, p. 857.
44 W.O. 3/289, p. 214.
45 *Journal of the Society for Army Historical Research*, XXVIII, 1950, 85.
46 Hugh Robinson's letters, 15 April and 4 June 1852.
47 Trial reports in *Military Review*, I, 1852, 151, 354; see also Rosa, *Colonel Colt, London*.
48 W.O. 46/86, p. 149.
49 P.P. 1854, XVIII, 159, 492.
50 W.O. 3/323, pp. 189, 194, 195.
51 W.O. 3/324, p. 284.
52 W.O.3/118, p.51; W.O. 3/326, p. 81.
53 Marquess of Anglesey (ed.), *Little Hodge*, p. 82; N.A.M. 6210/95, Estcourt to Wetherall, 18 December 1854, and 6210/97/24, Scarlett to Wetherall, 28 November 1855.
54 W.O. 3/325, pp. 161, 162, 444.
55 *Journal of the Society for Army Historical Research*, XIX, 1940, 12.
56 *Regulations for the Instruction, Formations and Movements of the Cavalry*, 1844, pp. 43–5.
57 *U.S.M.*, 1842, III, 210.
58 *U.S.J.*, 1831, II, 1; 1832, I, 289; 1833, III, 433; 1838, I, 38; Bunbury, *Reminiscences of a Veteran*, III, pp. 285–6.
59 *U.S.M.*, 1845, I, 115.
60 Smith, *System to Organize and Establish Military Discipline, in a Patriotic or Defensive Army*, pp. 139–40.
61 Nolan, *Cavalry: its History and Tactics*, pp. 292, 313.
62 Jones, *The Three Arms, or, Divisional Tactics, of Decker*, pp. 22, 33.
63 Marshman, *Memoirs of Major-General Sir Henry Havelock*, p. 182.
64 Nolan, *Cavalry: its History and Tactics*, p. 56.
65 Bismark, *Lectures on the Tactics of Cavalry*, pp. 330–1.

66 Nolan, *Cavalry: its History and Tactics*, p. v.
67 Luard, *History of the Dress of the British Soldier*, p. 135.
68 B.M. Add MSS 54561, ff 45–6, memo by C. Napier, 30 October 1840.
69 *U.S.J.*, 1835, I, 242, 398; 1835, II, 227.
70 Turner, *Practical Guide to the Breaking and Training of the Young Horse*, p. 23.
71 Sir George Cathcart papers, journal, Easter 1824.
72 W.O. 211/4, f 38; *British Army Despatch*, 21 March 1851, p. 187.
73 Sir George Cathcart papers, H 35, 37, 38, Cathcart to his father, 16 December 1829, 3 and 6 March 1830.
74 [Wathen?], *Remarks on the Proposed Regulations for the Instruction, Formations and Movements of the Cavalry; Spectateur Militaire*, 1st series, 36, 1843, 293.
75 Cathcart, *Correspondence of Lieut-General the Hon. Sir George Cathcart*, p. 398; Sir George Cathcart papers, M 41, Cathcart to his father, 12 August 1838.
76 Cecil Woodham-Smith, *The Reason Why*, p. 40.
77 *U.S.M.*, 1844, II, 438.
78 Charles Macfarlane, *The Camp of 1853*, p. 47; *U.S.G.*, 13 August 1853, 6.
79 Horse Guards circular, 13 July 1852; H. C. Wylly, *Military Memoirs of Lieut. General Sir Joseph Thackwell*, p. 354; W.O. 3/115, p. 252.
80 Nolan, *Cavalry: its History and Tactics*, pp. 143, 162–4.
81 Ibid., pp. 197, 279–82.
82 Sheppard, *Ninth Queen's Royal Lancers*, p. 125.
83 Anglesey (ed.), *Little Hodge*, p. 47.
84 B.M. Add MSS 31194, ff 36–9; also Sir George Cathcart papers, notebook on cavalry movements, c 1828.
85 Portland Collection, PwJf 2251, memo on cavalry, 1832; also 2281, orders for a review, 27 March 1833.
86 W.O. 135/3, memorandum (copy) 24 February 1844, para. 85; Von Bismark, *Lectures on the Tactics of Cavalry*, p. 197; *Aide-Mémoire to the Military Sciences*, II, p. 321.
87 Wellington papers, Russell to Wellington, 1 November 1833.
88 Sir George Cathcart papers, M 41, Cathcart to his father, 12 August 1838.
89 Kinloch in *Military Review*, II, 1853, 129; Kinloch, *Proposal for the Defence of the Country*, pp. 65–7; Bacon in *U.S.G.*, 26 August 1854, 6; 9 September 1854, 7.
90 *Military Review*, II, 1853, p. 193; Nolan, *Cavalry: its History and Tactics*, p. 322.
91 Wellington, *Despatches*, III, p. 353; N.A.M. 7709–6–13, p. 175, Vivian to Evans, 26 August 1835.
92 Le Couteur, *The Rifle*, pp. 28–9.
93 Nolan, *Cavalry: its History and Tactics*, pp. 78, 304.
94 Jervis, *Manual of Field Operations*, p. 6.
95 Wellington, *Despatches*, IV, p. 116.
96 P.P. 1851, VII, p. 11.
97 *U.S.J.*, 1841, I, 433.
98 Bismark, *Lectures on the Tactics of Cavalry*, p. 120.
99 Marquess of Anglesey (ed.), *Sergeant Pearman's Memoirs*, p. 85.
100 P.P. 1854–5, IX, *2nd Report*, p. 277.

101 Williams, *Historical Records of the Eleventh Hussars*, pp. 179, 230.
102 Newcastle papers, NeC 10485, Lucan to Newcastle, 10 June 1854.
103 Benson and Esher, *Letters to Queen Victoria*, III, p. 34.
104 N.A.M. 6807–281–1, Newcastle to Raglan, 10 April 1854.
105 Nolan, *Cavalry: its History and Tactics*, p. 64.
106 W.O. 3/115, p. 252.
107 Jervis, *Manual of Field Operations*, p. 5.
108 W.O. 3/112, pp. 136, 173.
109 Von Bismark, *Lectures on the Tactics of Cavalry*, p. 124.
110 Nolan, *Cavalry: its History and Tactics*, pp. 71–2.
111 Third Earl Grey papers, Charles Grey to Grey, 13 July 1841.
112 N.L.S. MSS 2845, ff 26–7, Brown to Arbuthnot, 14 March 1846.
113 Gordon-Brown (ed.), *Narrative of Private Buck Adams*, p. 36.
114 N.L.S. MSS 2846, f 18, Lt. Col. Michel to Brown, 24 January 1847.
115 N.L.S. MSS 2845, f 27, Brown to Arbuthnot, 14 March 1846.
116 Jervis, *Manual of Field Operations*, p. 7.
117 R.A. Add E/1, no. 155.
118 Luard, *History of the Dress of the British Soldier*, pp. viii, 150.
119 *Regulations for the Instruction, Formations and Movements of the Cavalry*, 1844, p. 1.
120 W.O. 3/283, p. 292.
121 *Queen's Regulations*, 1844, p. 371.
122 *U.S.M.*, 1844, II, 439.
123 T. Hookham, *Comparative View of the Form and Character of the English Race and Saddle-horse*, reviewed in *N. & M.G.*, 5 March 1836, 156–7.
124 P.P. 1854–5, IX, part III, *4th Report*, 49.
125 Robbins, *The Cavalry Catechism*, p. 170.
126 Wellington papers, Nicholls to Wellington, 16 December 1842.
127 Liddell, *Memoirs of the Tenth Royal Hussars*, p. 230.
128 Sir George Cathcart papers, Cathcart to Airey, 13 January 1853.
129 India Office Library, L/MIL/3/1910, no. 9.
130 Nolan, *Cavalry: its History and Tactics*, pp. 333–6; see also *On the Deteriorated Condition of our Saddle-horses*, pp. 2, 55.
131 *U.S.M.*, 1855, I, 258; P.P. 1854–5, IX, part II, *3rd Report*, 336; Wrottesley (ed.), *Military Opinions of Burgoyne*, pp. 417–18.
132 McClellan, *Report of Captain George B. McClellan ... One of the Officers Sent to the Seat of War in Europe in 1855 and 1856*, p. 271.
133 N.L.S. Adv MSS 46.9.5, f 70, Somerset to Murray, 21 January 1843.
134 W.O. 3/112, p. 471.
135 R.A. Add E/1, no. 155.
136 P.P. 1854–5, IX, part III, *4th Report*, p. 249; W.O. 3/115, p. 252.
137 P.P. 1856, XXI, p. 426.
138 *U.S.G.*, 4 June 1853, 4.
139 Robbins, *The Cavalry Catechism*, pp. 54–5.
140 Hardinge, *Rulers of India: Viscount Hardinge*, p. 84.
141 Anglesey (ed.), *Sergeant Pearman's Memoirs*, p. 70.
142 Lt. Col. von Arentschildt, *Instructions for Officers and Non-commissioned Officers of Cavalry, on Outpost Duty*, pp. 30–3.
143 B.M. Add MSS 31194, f 27, Von Muller, 'Treatise ... for Cavalry', 1836.
144 Verner, *Military Life of Duke of Cambridge*, I, pp. 5–6; W.O. 44/677, p. 746;

Raglan private papers, box 3, no. 126, Somerset to Wellington, 16 June 1832; *Journal of the Society for Army Historical Research*, XXVIII, 1950, 139.
145 W.O. 3/91, p. 46; 3/177, p. 371; 3/393, p. 70; N.A.M. 7709–6–13, p. 289, Howick to Vivian, 29 June 1837 and Vivian to regiments, 2 July 1837.
146 Anglesey, *History of the British Cavalry*, I, p. 261. A contemporary drawing, reproduced as plate 30, does not appear to show carbines.
147 N.L.S. MSS 1859, ff 22–3, Brown to Brotherton, 10 February 1852; W.O. 3/112, p. 256; 3/113, p. 94; 3/115, pp. 140, 184; 3/315, pp. 183, 209, 212, 291; 3/316, p. 348; 3/318, p. 32.
148 W.O. 44/677, pp. 864–749.
149 W.O. 44/677, pp. 973, 969, 743–702, 536, 507, 405–304; W.O. 3/289, p. 149.
150 W.O. 3/297, p. 188.
151 W.O. 3/110, p. 114.
152 W.O. 33/1, p. 621.
153 W.O. 3/293, p. 60; 3/294, p. 335; 44/539.
154 Roads, *The British Soldier's Firearm 1850–1864*, pp. 43–4.
155 W.O. 3/320, p. 18.
156 W.O. 3/325, p. 315.
157 W.O. 3/323, pp. 74, 126, 143, 441.
158 W.O. 33/1, p. 621.
159 W.O. 3/326, p. 33.
160 W.O. 3/324, p. 310.
161 *Instructions for the Carbine Exercise*, 1844 edn, p. 15.
162 *Regulations for the Instruction, Formations, and Movements of the Cavalry*, 1833, pp. 101–2.
163 *Naval and Military Magazine*, III, 1828, xxii.
164 *Royal Military Magazine*, I, 1852, 335.
165 *Journal of the Society of Army Historical Research*, XXII, 1944, p. 29; XXIII, 1945, p. 41; Bisset, *Sport and War*, pp. 120, 135–6.
166 Lt. Col. T. W. Taylor in *U.S.J.*, 1835, II, 227.
167 Knollys (ed.), *Life of General Sir Hope Grant*, I, p. 143.
168 *U.S.J.*, 1831, II, 69; 1834, III, 399; *N. & M.G.*, 27 May 1843, 326; 18 November 1843, 726.
169 Bismark, *Lectures on the Tactics of Cavalry*, pp. 60–1.
170 Nolan, *Cavalry: its History and Tactics*, pp. 65–6.
171 Wellington, *Despatches*, I, p. 102.
172 Alexander, *Passages in the Life of a Soldier*, I, pp. 136–8; *U.S.M.*, 1854, III, 1.
173 *British Army Despatch*, 15 April 1853, 243.
174 *U.S.M.*, 1842, I, 392–3; III, pp. 355–6; Alexander, *L'Acadie*, I, p. 44; *U.S.G*, 3 March 1838, 6.
175 W.O. 3/92, p. 118.
176 Sir George Cathcart papers, M 103, 'Instructions for cavalry in Canada', 1840; also notebook on cavalry movements c 1828, and memo on Canada, c 1845.
177 N.L.S. Adv MSS 46.9.5, f 70, Somerset to Murray, 21 January 1843; W.O. 3/296, pp. 339, 428; W.O. 3/99, p. 281.
178 W.O.3/106, p. 305.
179 W.O. 3/109, p. 481.
180 W.O. 43/707, ff 165–74.

181 N.L.S. Adv MSS 46.9.6, f 27, G. Napier to Stanley, 16 December 1842.
182 *U.S.M.*, 1847, II, 284; N.L.S. MSS 2846, f 19, Michel to Brown, 24 January 1847; E. Elers Napier, *Excursions in Southern Africa*,. II, p. 404.
183 W.O. 3/311, p. 41; Third Earl Grey papers, Smith to Grey, 26 April 1849, 25 May 1849, 26 March 1850.
184 Hugh Robinson letter of 10 April 1852.
185 King, *Campaigning in Kaffirland*, pp. 188, 213; *U.S.G.*, 10 April 1852, 3; Hardinge papers, J. Bruce Pryce to Hardinge, 25 November 1852.
186 W.O. 3/291, p. 145; 3/105, p. 66.
187 W.O. 3/112, pp. 136, 173, 256; 3/113, p. 94; 3/115, pp. 140, 184; 3/315, pp. 183, 209, 212, 291; 3/316, p. 348; 3/318, p. 32; N.L.S. MSS 1859, ff 22–3, Brown to Brotherton, 10 February 1852.
188 Roads, *The British Soldier's Firearm 1850–1864*, pp. 42–3.
189 N.L.S. MSS 2846, f 18, Michel to Brown, 24 January 1847.
190 S.R.O. GD 45/8/87, H. Somerset to Harry Smith, 24 February 1851.
191 F.R. Chesney, *Observations on the Past and Present State of Fire-arms*, pp. 293–4, 304–5.
192 Quoted in Douglas, *A Treatise on Naval Gunnery*, pp. 351–2.
193 Herbert papers, Palmerston to Herbert, 21 September 1853, *re* Devon Yeomanry.
194 General Yusuf in *U.S.M.*, 1851, III, 418–19.
195 *U.S.J.*, 1831, II, 205.
196 Bruce (ed.), *Life of General Sir William Napier*, II, p. 396.

### 4 Artillery: the arm of the future

1 [Simmons]. *Remarks on the Promotion of the Officers of the Corps of Artillery in the British Service, and on the Application of that Arm in the Field*, pp. 1, 6.
2 W.O. 3/97, p. 149.
3 Benson and Esher, *Letters of Queen Victoria*, II, pp. 239–40.
4 Third Earl Grey papers, Russell to Grey, 21 October 1848.
5 N.L.S. MSS 9870, f 57, Pasley's diary, 16 and 17 July 1849.
6 N.A.M. 7709–6–14, pp. 90–1, Vivian to Dickson, 19 September 1838.
7 W.O. 44/618; P.P. 1849, IX, 111–13.
8 W.O. 33/1, p. 106.
9 Hime, *History of the Royal Regiment of Artillery*, p. 66.
10 *Quarterly Review*, 82, 1848, 471–2.
11 *U.S.M.* 1842, III, 375.
12 P.P. 1849, IX, 374; Chesney, *Observations on the Past and Present State of Fire-arms*, pp. 142–5, 154.
13 Wellington papers, Chesney to Wellington, 20 January 1840, and reply.
14 W.O. 30/124, pp. 10–11.
15 P.P. 1849, IX, 386–7.
16 *Aide-Mémoire to the Military Sciences*, I, p. 437.
17 Riddell, *Remarks on the Organization of the British Royal Artillery*, pp. 22–4.
18 P.P. 1840, XXII, 53.
19 W.O. 3/98, pp. 21, 160; W.O. 3/294, p. 404; N.L.S. Adv MSS 46.9.13, pp. 2–4, Murray to Board of Ordnance, 3 January 1842.
20 W.O. 3/104, p. 27; W.O. 3/308, p. 40; W.O. 44/542; P.P. 1849, IX, 24.
21 W.O. 44/539.

22 Chesney, *Observations on the Past and Present State of Fire-arms*, p. 130.
23 Hime, *History of the Royal Regiment of Artillery*, p. 18.
24 *Aide-Mémoire to the Military Sciences*, I, pp. 434–5.
25 W.O. 44/644.
26 P.P. 1840, XXII, 56.
27 W.O. 44/644, Jones to Paget, 24 September 1847.
28 R.A. Add E/1, no. 175, p. 12, E. C. Warde, 'The Wants and Defects of the Regiment of Artillery', 1849.
29 Gardiner, *Illustrations ... of the Artillery of the British Army*, p. 19.
30 P.P. 1849, IX, xii.
31 *U.S.G.*, 6 March 1847, 7.
32 *N. & M.G.*, 15 July 1848, 458.
33 Ibid., 12 October 1850, 645.
34 Ibid., 25 April 1846, 260.
35 *U.S.M.*, 1852, II, 183–6, 329.
36 *N. & M.G.*, 9 July 1853, 436.
37 *U.S.J.*, 1831, II, 43.
38 An Officer of the Corps, *The Royal Artillery: a Few Remarks on its Deficiency and Inefficiency*, p. 17.
39 *N. & M.G.*, 27 February 1847, 139; 15 July 1848, 458.
40 Ibid., 4 January 1834, 13.
41 Gardiner, *Report on the ... Artillery of the British Army*, p. 14.
42 Evidence to 1828 Finance Committee, quoted in *U.S.J.*, 1831, I, 382.
43 P.P. 1849, IX, 652–3.
44 N.L.S. Adv Mss 46.9.14, pp. 101–2, Murray to Goulburn, 31 January 1843.
45 Hansard, 3rd series, XLII, c 625.
46 B.M. Add MSS 40582, f 53, Murray to Peel, 3 January 1846.
47 Hansard, 3rd series, XLII, c 632.
48 S.R.O., GD 45/8/87, despatch of 18 February 1851.
49 W.O. 44/644, Ross to Paget, 8 September 1847.
50 R.A. E 42/51, pp. 3–4, Gardiner to Russell, 26 January 1848.
51 P.P. 1849, IX, 652–3.
52 W.O. 44/510, Director General of Artillery's return for 6 February 1854.
53 Chesney, *Observations on the Past and Present State of Fire-arms*, p. 87.
54 Head, *Defenceless State of Great Britain*, p. 64.
55 Wellington papers, Kempt to Wellington, 1 January 1833, enclosing Dickson's memorandum of 28 February 1832.
56 Whinyates, *From Coruña to Sebastopol*, p. 293.
57 Gardiner, *Report on the ... Artillery of the British Army*, p. 7; R.A. E 42/35, p. 9, Gardiner to Russell, 15–16 June 1848.
58 P.P. 1849, IX, 54–5.
59 W.O. 6/127, p. 32; N.A.M. 7709–6–14, p. 65, Vivian to Glenelg, 18 April 1838; p. 100, Vivian to Melbourne, 3 December 1838; p. 103, Vivian to Rice, 2 February 1839.
60 W.O. 44/541; Hansard, 3rd series, XLII, c 625, and LVII, c 36.
61 N.L.S. Adv MSS 46.9.15, p. 120, Murray to Peel, 13 August 1844; W.O. 80/3, Murray to Peel, 25 August 1844.
62 B.M. Add MSS 40461, f 277, Wellington to Murray, 5 October 1845.
63 Ibid., 40582, ff 52–8, Murray to Peel, 3 January 1846.

64 P.P. 1849, IX, ix, x, 20, 24–5; Hansard, 3rd series, XC, c 980; W.O. 44/522; R.A. B 10/219.
65 *U.S.G.*, 6 May 1848, 3.
66 Hardinge papers, Elliot to Hardinge, 29 October 1852, with enclosures from Ross.
67 *N. & M.G.*, 18 June 1853, 392.
68 R.A. E 42/35, Gardiner to Russell, 15–16 January 1848; R.A. E 42/51, 26 January 1848.
69 Gardiner, *Report on the ... Artillery of the British Army*, pp. 28–9.
70 W.O. 46/84, p. 326; W.O. 6/128, p. 174; W.O. 1/601, pp. 415–17.
71 R.A. E 44/39, memo by Hardinge, 22 November 1852.
72 R.A. E 44/15, memo by Hardinge, 23 September 1852.
73 R.A. E 44/39, memo by Hardinge, 22 November 1852.
74 R.A. E 44/42, memo by Raglan, 24 November 1852.
75 W.O. 6/128, p. 215; W.O. 1/601, pp. 423–5; Hansard, 3rd series, CXXII, c 1017.
76 Third Earl Grey papers, Charles Grey to Grey, 28 November 1852.
77 W.O. 1/601, p. 225.
78 Hime, *History of the Royal Regiment of Artillery*, p. 87.
79 Archives Historiques de Guerre, MR 1430/15.
80 Second Earl Cathcart's papers, J 42, Raglan to Cathcart, 20 January 1853; W.O. 46/87, pp. 150, 289; *Military Review*, II, 1853, 55.
81 P.P. 1854–5, IX, part 2, *3rd Report*, 335.
82 R.A. E 44/15, p. 30.
83 *N. & M.G.*, 18 January 1834, 43.
84 R.A. Add E/1, no. 175, pp. 7–8, E. C. Warde, 'The Wants and Defects of the Regiment of Artillery', 1849; Gardiner, *Illustrations ... of the Artillery of the British Army*, p. 20; P.P. 1849, IX, 52.
85 P.P. 1849, IX, 53; *U.S.M.*, 1849, I, 586; *N. & M.G.*, 26 August 1848, 555, and 2 September 1848, 571.
86 P.P. 1849, IX, 36.
87 N.A.M. 6807/289, Ross to Raglan, 11 August 1854.
88 *N. & M.G.*, 19 June 1847, 376.
89 Ibid., 18 September 1847, 601, and 2 October 1847, 632; Mazé's observations in G.-A. Jacobi, *Etat actuel de l'artillerie de campagne en Europe: artillerie de campagne anglaise*, p. 154; R. A. Add E/1 no. 175, p. 14; P.P. 1849, IX, 388–9.
90 *N. & M.G.*, 26 February 1848, 130.
91 Gardiner, *Illustrations ... of the Artillery of the British Army*, p. 14.
92 Eardley-Wilmot, *Memorials of Fredk. M. Eardley-Wilmot*, pp. 81–2, 114.
93 R. A. E 44/15, pp. 43–4.
94 Spearman, *Notes on Military Education*, p. 76.
95 *N. & M.G.*, 26 August 1848, 555.
96 *U.S.G.*, 23 November 1850, 3.
97 An Officer of the Corps, *The Royal Artillery: a Few Remarks on its Deficiency and Inefficiency*, p. 7.
98 P.P. 1849, IX, 264.
99 W.O. 46/90, 389.
100 Jocelyn, *History of the Royal Artillery (Crimean Period)*, p. 18.

101 *Remarks on the Education of the Royal Artillery* [possibly by Eardley-Wilmot], p. 13.
102 W.O. 44/298.
103 P.P. 1849, IX, 456, 608.
104 *N. & M.G.*, 3 July 1841, 420; 10 July 1841, 434; 17 July 1841, 451; 24 July 1841, 466; 7 August 1841, 500.
105 B.M. Add MSS 41581, f 145, report by Dansey, 4 May 1846.
106 W.O. 44/298.
107 B.M. Add MSS 41581, ff 142–8, report by Dansey, 4 May 1846.
108 W.O. 44/618.
109 W.O. 44/644.
110 W.O. 44/63; W.O. 44/125.
111 *N. & M.G.*, 30 June 1849, 402; Connolly, *History of the Royal Sappers and Miners*, II, p. 40.
112 *N. & M.G.*, 14 July 1849, 436; *British Army Despatch*, 29 June 1849, 807.
113 W.O. 46/89, pp. 118–19, 192–3.
114 Hogg, *The Royal Arsenal*, I, p. 668.
115 Straith, *Treatise on Fortification and Artillery*, II, p. 109.
116 B.M. Add MSS 39998, f 12, C. Dickson, 'A Practical Course of Artillery', 1835.
117 *U.S.M.*, 1845, I, 373; Griffiths, *Artillerist's Manual*, 1839, p. 42.
118 Boxer, *Treatise on Artillery*, pp. 55–71, 173.
119 *U.S.J.*, 1839, III, 381 gives .06 inch as Dickson's achievement, but this must be exceptional; B.M. Add MSS 45897, f 12, quotes .1 inch as the normal windage in brass field guns in 1845, and .15 inch to .2 inch in iron; those given above are from Hughes, *British Smooth-Bore Artillery*, p. 21, and are confirmed in *Minutes of Proceedings of Royal Artillery Institution*, I, 1857, 24–30.
120 P.P. 1849, IX, 164.
121 B.M. Add MSS 41581, f 143, report by Dansey, 4 May 1846.
122 W.O. 44/541.
123 W.O. 186/1, p. 37; W.O. 46/87, p. 133; W.O. 46/90, p. 201.
124 *U.S.J.*, 1841, II, 272.
125 B.M. Add MSS 41581, f 145, report by Dansey, 4 May 1846.
126 Lefroy (ed.), *Handbook for Field Service*, pp. 8, 10.
127 Ibid., p. 4; these figures do not agree exactly with the table on pp. 2–3.
128 B.M. Add MSS 41581, f 64, memo by Dickson, 8 September 1839.
129 *Aide-Mémoire to the Military Sciences*, III, pp. 405–6.
130 Hughes, *Firepower*, p. 38.
131 Thackwell, *Narrative of the Second Seikh War*, p. 92.
132 Lefroy, *Handbook for Field Service*, p. 4; W.O. 44/618.
133 W.O. 186/1, pp. 27, 49–53, 77.
134 W.O. 46/87, p. 353.
135 W.O. 46/91, p. 273.
136 Straith, *Treatise on Fortification and Artillery*, I, p. 15.
137 Oakes, *The Young Artillery Officer's Assistant*, appendix, pp. 2–3.
138 *Aide-Mémoire to the Military Sciences*, III, p. 109.
139 Hughes, *Firepower*, pp. 37–8, 65.
140 B.M. Add MSS 39998, f 8, C. Dickson, 'A Practical Course of Artillery', 1835.
141 B.M. Add MSS 45897, f 4, E. H. East, 'A Course of Practical Artillery', 1845.

142 Jervis, *Manual of Field Operations*, p. 111.
143 Whinyates, *From Coruña to Sebastopol*, p. 137.
144 *U.S.J.*, 1841, II, 273.
145 *U.S.M.*, 1852, II, 212; Archives Historiques de Guerre, MR 1425/74, p. 28.
146 *Aide-Mémoire to the Military Sciences*, I, p. 204.
147 *N. & M.G.*, 25 August 1849, 530.
148 B.M. Add MSS 45897, ff 62–73; Jacobi, *Etat actuel de l'artillerie de campagne en Europe: artillerie de campagne anglaise*, pp. 130–1; *U.S.M.*, 1849, II, 1–10.
149 Dupin, *Military Force of Great Britain*, II, pp. 224–6.
150 *U.S.M.*, 1849, II, 198.
151 B.M. Add MSS 49107, ff 84–5, 92, C. Napier to Robertson, 20 July 1845, 17 April 1847; Add MSS 49115, ff 122–9, unsigned memo; Add MSS 54536, f 78, Robertson to C. Napier, 5 March 1847; *N. & M.G.*, 2 January 1847, 8; 6 February 1847, 90; 13 February 1847, 105; 20 February 1847, 121; 6 March 1847, 154–5; 6 December 1851, 778.
152 B.M. Add MSS 49107, f 92, C. Napier to Robertson, 17 April 1847.
153 P.P. 1849, IX, 379.
154 *N. & M.G.*, 4 November 1848, 715.
155 Rait, *Life of Gough*, I, pp. 371, 374.
156 P.P. 1849, IX, 606–7; *N. & M.G.*, 27 May 1843, 323; 10 June 1848, 372; 3 March 1849, 139; *Minutes of Proceedings of the Royal Artillery Institution*, I, 1857, 32.
157 Hogg, *The Royal Arsenal*, p. 686; *N. & M.G.*, 3 February 1849, 66.
158 *N. & M.G.*, 15 October 1853, 663; *U.S.G.*, 15 October 1853, 3.
159 R.A. E 5/26, Stovin to Grey, 25 January [1855].
160 National Register of Archives report no. 12702, 'Crimean Letters of Colonel David Wood 1854–1856', nos. 63, 236, 237.
161 Wellington papers, Kempt to Wellington, 1 January 1833, enclosing Dickson's memorandum of 28 February 1832.
162 Ibid., Wellington to Kempt, 4 January 1833, and Wellington to Anglesey, 1 October 1846 and 13 April 1847; P.P. 1849, IX, 326.
163 *The War in India*, pp. 61, 117; Wellington papers, Charles Abbott to ? Wellington, 19 January 1846.
164 Straith, *Treatise on Fortification and Artillery*, II, pp. 137–8, 228.
165 *Royal Engineers Professional Papers*, new series, IV, 1840, 91.
166 Chesney, *Observations on the Past and Present State of Fire-arms*, pp. 194, 320–41; see also *U.S.M.*, 1853, II, 576; Favé, *Emperor Napoleon's New System of Field Artillery*.
167 N.A.M. 6807/289, Raglan to Ross, 13 April 1854, and replies, 14 December 1854, 22 December 1854, 1 January 1855.
168 Jocelyn, *The History of the Royal Artillery (Crimean Period)*, pp. 131, 256, 261–8.
169 P.P. 1849, IX, 325; Thackwell, *Narrative of the Second Seikh War*, p. 253.
170 Hardinge papers, Hardinge to Prince Albert, 29 August 1855.
171 Fullom, *Life of General Sir Howard Douglas*, pp. 379–82; B.M. Add MSS 40475, ff 186, 192–3, Hardinge to Peel, 31 March 1846.
172 Shadwell, *Life of Colin Campbell, Lord Clyde*, I, p. 221.
173 Hansard, 3rd series, CX, c 44.
174 *U.S.M.*, 1847, II, 346.

175 Yates, *Elementary Treatise on Tactics*, p. 145.
176 *U.S.J.*, 1831, I, 65.
177 *U.S.M.*, 1847, I, 481, and II, 346.
178 *N. & M.G.*, 15 January 1848, 41.
179 Hansard, 3rd series, CXXIII, c 109.
180 Straith, *Treatise on Fortification and Artillery*, I, p. viii.
181 William Malins, *A Plan for Additional National Defence*, p. 30.
182 *Royal Engineers Professional Papers*, VII, 1845, 8.
183 P.P. 1849, IX, xxix, 179, 186.
184 Douglas, *A Treatise on Naval Gunnery*, p. 225; *Edinburgh Review*, XCVI, 1852, 512.
185 Calculated as the squares of the diameter of shot, Straith, *Treatise on Fortification and Artillery*, II, p. 228.
186 N.A.M. 7709–6–14, pp. 81–5, Vivian to Dickson, 15 July 1838.
187 The table is printed in the *Aide-Mémoire to the Military Sciences*, I, p. 60, and is also to be found in W.O. 55/1409, pp. 118–19.
188 P.P. 1849, IX, 971.
189 W.O. 46/89, p. 37.
190 *N. & M.G.*, 6 July 1839, 422; Scoffern, *Projectile Weapons of War*, pp. 187–8; Faddy, *Essay on the Defence and Military System of Great Britain*, p. 10.
191 W.O. 55/1563/7, pp. 1–17.
192 W.O. 55/1409, p. 10; for trials, see *N. & M.G.*, 4 June 1842, 360, and 3 June 1843, 338.
193 R.A. E 44/59, Hardinge to Albert, 29 November 1852; Hansard, 3rd series, CXXIII, c 1017.
194 *Aide-Mémoire to the Military Sciences*, I, pp. 48–9, 443.
195 Wellington papers, Dickson memorandum of 28 February 1832, enclosed by Kempt to Wellington, 1 January 1833.
196 H. Sandham papers, memo book for 1846, notes on mortars inserted in book, 1849–50.
197 B.M. Add MSS 45897, f 21, E. H. East, 'A Course of Practical Artillery', 1845.
198 *Royal Engineers Professional Papers*, II, 1838, 49.
199 Jocelyn, *History of the Royal Artillery (Crimean Period)*, pp. 176–7.
200 B.M. Add MSS 54536, ff 39, 45, 61, 72, Robertson to C. Napier, 14 September 1842, 29 October 1842, 26 June 1843, 30 August 1845.
201 Straith, *Treatise on Fortification and Artillery*, II, pp. 34–8; Wellington papers, Egerton to Wellington, 2 September 1845; Humfrey, *An Essay on the Modern System of Fortification*, pp. 10–11.
202 Wellington papers, Wellington to Murray, 23 October 1845.
203 Maurice [de Sellon], *On National Defence in England*, pp. 78–80.
204 *Military Review*, I, 1852, 346; *U.S.J.*, 1830, II, 831.
205 *Aide-Mémoire to the Military Sciences*, I, pp. 56–61.
206 Quoted in An Artillery Officer, *Considerations on Tactics*, p. 60.
207 W.O. 30/118, C. J. Napier to William Napier, *c.* 1843.
208 B.M. Add MSS 40475, f 128, Hardinge to Queen, 18 February 1846.
209 [Ryder], *Four Years' Service in India*, p. 62; *Aide-Mémoire to the Military Sciences*, III, pp. 408–21.
210 *U.S.M.*, 1853, I, 292.

211 Wellington papers, Murray to Wellington, 8 November 1845.
212 Hansard, 3rd series, CVII, c 544.
213 C. J. Napier, *Defects, Civil and Military, of the Indian Government*, p. 397.
214 Straith, *Treatise on Fortification and Artillery*, I, p. 24; II, p. 49.
215 *U.S.J.*, 1831, I, 87.
216 *N. & M.G.*, 16 August 1834, 658.
217 James Fergusson, *An Essay on a Proposed New System of Fortification.*
218 *U.S.M.*, 1851, II, 257; 1852, II, 429.
219 *Military Review*, II, 1853, 209, 277; N.L.S. MSS 9872, ff 186–205, Pasley's diary, 20 April–29 May 1853.
220 Jebb, *A Flying Shot at Fergusson*, p. 32; *U.S.M.*, 1853, I, 522.
221 Seaton papers, Yonge to Seaton, n.d. [? May 1853].
222 P.P. 1849, IX, xiv, xvi, 68.
223 Jocelyn, *History of the Royal Artillery (Crimean Period)*, p. 457.
224 Mordecai, *Military Commission to Europe, in 1855 and 1856*, p. 66.
225 Delafield, *Report on the Art of War in Europe in 1854, 1855, and 1856*, p. 55.
226 N.A.M. 6210–97–20, Lysons to Wetherall, 16 November 1855.
227 Wrottesley, *Life of Burgoyne*, II, p. 152.
228 Porter, *History of the Corps of Royal Engineers*, I, p. 458.
229 Jocelyn, *History of the Royal Artillery (Crimean Period)*, p. 457.
230 Henry Wilkinson, *Engines of War*, p. 122.
231 *U.S.J.*, 1841, I, 371.
232 W. F. P. Napier, *C. J. Napier*, II, p. 193.
233 Hughes, *Bengal Horse Artillery*, p. 59.
234 W.O. 44/618; B.M. Add MSS 41581, f 136, memoir on rockets, July 1844.
235 W.O. 46/89, pp. 188, 292.
236 Hogg, *The Royal Arsenal*, p. 751.
237 B.M. Add MSS 45897, ff 118–20, E. H. East, 'A Course of Practical Artillery', 1845.
238 P.P. 1849, IX, 53.
239 *U.S.M.*, 1854, I, 606; *N. & M.G.*, 8 April 1854, 217.
240 W.O. 44/508; see also Douglas, *A Treatise on Naval Gunnery*, pp. 245–53; *Military Review*, I, 1852, 82–4; Boxer, *Treatise on Artillery*, p. 174.
241 *N. & M.G.*, 7 February 1852, 90; 28 February 1852, 135; 20 March 1852, 181; 7 August 1852, 501; W.O. 46/86, 7–8, 36, 38, 161; *Military Review*, I, 1852, 82–4.
242 Douglas, *A Treatise on Naval Gunnery*, p. 609.
243 W.O. 46/89, pp. 199, 261, 297.
244 Hardinge papers, Foreign Office to Secretary to the Ordnance, 16 June 1852.
245 W.O. 46/89, pp. 222, 260, 261.
246 Mordecai, *Military Commission to Europe, in 1855 and 1856*, p. 110; W.O. 44/628, quoted in an unpublished paper by C. I. Hamilton entitled 'The Lancaster Gun'.
247 A gradual increase in this order can be traced. Graham papers, Raglan to Graham, 25 March 1854, says six 8-inch guns of sixty-five hundredweight and two of ninety-five hundredweight prepared for Royal Navy. Newcastle papers, NeC 10597, report by Col. Chalmer, 8 May 1854, says five 8-inch guns of ninety-five hundredweight, seven of sixty-five hundredweight, seven of fifty-two hundredweight have been bored on Lancaster's system, and four of fifty-two hundredweight sent to the East.

248 Wrottesley, *Life of Burgoyne*, II, p. 107; N.A.M. 6807–283, Raglan to Newcastle, 18 and 29 October 1854.
249 H. Sandham papers, memo book for 1846, pp. 40–6, 24 May–8 December 1854.
250 W.O. 44/298.
251 W.O. 186/1, pp. 124–5.
252 Mordecai, *Military Commission to Europe, in 1855 and 1856*, pp. 110–11.

**Conclusion**

1 Shaw (ed.), *Great Britain and the Colonies*, p. 2.
2 Griffith, 'Military Thought in the French Army'; Samuel P. Huntington, *The Soldier and the State*, Cambridge, Mass., 1957, pp. 217–21.
3 Quoted in Jay Luvaas, *The Military Legacy of the Civil War: The European Inheritance*, Chicago, 1959, p. 185.

# Bibliography

The bibliography is arranged as follows:

I. Unpublished papers
II. Parliamentary papers
III. Periodicals
IV. Contemporary published works
V. Diaries, memoirs, etc.
VI. Secondary works

### I. Unpublished papers


Archives Historiques de Guerre, Château de Vincennes, Paris, reports on the British Army

Bodleian Library, Oxford
- Clarendon Deposit
- Napier papers (predominantly those of William)
- North MSS, papers of Colonel J. S. Doyle (later North)

British Museum
- Add MSS 31194, Von Muller, 'Treatise on Field Service for Cavalry', 1836
- Add MSS 35060, Hill papers
- Add MSS 37415, Wellesley papers
- Add MSS 39998, C. Dickson, 'A Practical Course of Artillery', 1835
- Add MSS 40310, 40313–14, 40459–61, 40474–5, 40575, 40581–5, Peel papers, especially correspondence with Hardinge and Wellington
- Add MSS 41581, Col. Dansey papers
- Add MSS 41964, Sir C. W. Pasley papers
- Add MSS 45897, E. H. East 'A Course of Practical Artillery', 1845
- Add MSS 49105–8, 49115–17, 49128–31, 49140, 49143, 49144, 49168, 54512–19, 54524–6, 54532, 54536, 54538, 54544–61, Napier papers (predominantly those of Charles)
- Add MSS 49480, Sir J. W. Gordon papers

The Earl Cathcart
- Second Earl Cathcart papers
- Sir George Cathcart papers

Dr A. E. Clark-Kennedy
- Transcripts of Colonel John Clark-Kennedy's papers

Durham University, Department of Palaeography and Diplomatic
- Sir Charles Grey papers
- Third Earl Grey papers

Greater London Record Office (Middlesex Records)
Colonel David Wood's letters, 1854–6 (see National Register of Archives report no. 12702)
Hereford County Record Office
Airey papers (see National Register of Archives report no. 10872)
R. E. Howard-Vyse, Esq.
Hugh Robinson's letters, 1851–3
India Office Library
L/MIL/3/64, 1752–4, 1905, 1907–10. Reports concerning Lord Frederick Fitzclarence at Bombay
J. E. Colborne Mackrell, Esq.
Seaton papers
McGill University Library
Hardinge papers
Ministry of Defence Whitehall Library
General Orders, Horse Guards circulars and memoranda
National Army Museum
A. S. H. Mountain papers
Raglan military papers (relates to Crimea only)
Sir Hussey Vivian papers
Sir George Wetherall papers
National Library of Scotland
MSS 1847–50, 1855, 1857, 1859, 2839–54, 3259–62, Sir George Brown papers
MSS 3867, General Walter Scott papers
MSS 3870, G. R. Gleig papers
MSS 8105, journal of Captain C. H. Scott-Douglas
MSS 9319, 'Crimean Campaign 1854–1856', Scots Fusilier Guards
MSS 9869–74, 9880–2, Sir C. W. Pasley papers
Adv MSS 46.8.16–23, 46.9.1–19, Sir George Murray papers
Nottingham University Library
Fifth Duke of Newcastle papers
Portland Collection, letters to Lord William Bentinck from Lt. Gen. Sir S. F. Whittingham
Earl of Pembroke
Sidney Herbert papers
Public Record Office
P.R.O. 30/46, Sir William Eyre papers
W.O. 1/595–601, in-letters, Colonial Department
W.O. 3/90–119, 283–329, Adjutant-General's out-letters
W.O. 4/263–72, Secretary at War's out-letters
W.O. 6/127–8, Secretary of State's out-letters
W.O. 7/57–9, Reports of the Board of General Officers
W.O. 30/112, 116, 118, miscellaneous papers (? Commander-in-Chief's)
W.O. 33/1, War Office confidential papers
W.O. 37, Sir George Scovell papers
W.O. 43, Secretary at War, 'Very Old Series' and 'Old Series'
W.O. 44, Board of Ordnance in-letters
W.O. 46/86–92, Master General of the Ordnance out-letters
W.O. 55/970 and 1931, Royal Engineers' instruction
W.O. 55/1409 and 1563/7, coast defence

W.O. 80, Sir George Murray papers
W.O. 99, Royal Military College, Sandhurst
W.O. 135, Sir Harry Smith papers
W.O. 186/1, reports from Shoeburyness for Director General of Artillery
W.O. 211, Hart papers on officers' biographies
Lord Raglan
Raglan private papers
Royal Archives
Duke of Cambridge papers
Queen Victoria papers
Royal Engineers Museum, Chatham
Sir John Burgoyne papers
Colonel H. Sandham memorandum books
Scottish Record Office
GD 24/1/936, Cornet Drummond papers
GD 24/8/1–24, Panmure papers
GD 26/9/546, Ensign H. F. B. Maxse's memorandum book
GD 225/boxes 41–2, Leith Hay papers
Duke of Wellington
First Duke of Wellington papers
East Yorkshire County Record Office
DDCH 76–84, 118–19, Sir Charles Chichester papers
DDCH 95–110, C. R. Chichester's diaries

**II. Parliamentary papers**

P.P. 1849, IX, P.P. 1850, X, and P.P. 1851, VII, p. 735, Report from the select committee on army and Ordnance expenditure
P.P. 1854, XVIII, Report from the select committee on small arms
P.P. 1854–5, IX, 1st–5th Reports from the select committee on the army before Sebastopol
P.P. 1856, XX, Report of the commission of inquiry into the supplies of the British army in the Crimea
P.P. 1856, XXI, Report of board of general officers appointed to inquire into the statements contained in the reports of Sir John McNeill and Colonel Tulloch
Hansard's Parliamentary Debates, 3rd series, vols. XXVII–CXXXV

**III. Periodicals**

*The Army Reformer*
*Blackwood's Magazine*
*British Army Despatch*
*Edinburgh Review*
*Fraser's Magazine*
*Journal des Sciences Militaires*
*Military Annual for 1844*
*Military Magazine*
*Military Review*
*Minutes of Proceedings of the Royal Artillery Institution*
*Naval and Military Gazette*

*Naval and Military Magazine* (which became the *United Service Journal*)
*Quarterly Review*
*Royal Engineers Professional Papers* (or *Papers on subjects connected with the Duties of the Corps of Royal Engineers*)
*Royal Military Magazine* (at first called the *British Soldier*)
*Le Spectateur Militaire*
*United Service Gazette*
*United Service Journal*, which became the *United Service Magazine*
*Westminster Review* (called *London and Westminster Review* 1836–40)

### IV. Contemporary published works

Official manuals are grouped under the headings 'Horse Guards' and 'Ordnance, Board of' according to the issuing authority. Anonymous works are listed under the first word of the title; pseudonymous under the pseudonym.

Adye, Capt. R. W., *The Bombardier and Pocket Gunner*, London, 1827
*Aide-Mémoire to the Military Sciences*, 3 vols., London, 1846–52
Ainslie, Maj. C. P., *The Cavalry Manual*, London, 1843
Anderson, James, *System of National Defence*, Edinburgh, 1853
Angelo, Col. [E.A.], *Observations on the Military Exercises Formed by (the late) Henry Angelo, Esq.*, London, 1853
[Angelo, Henry], *Angelo's Bayonet Exercise*, London, 1853
Angelo, Henry, *Instructions for the Sword Exercise: Selected from His Majesty's Rules and Regulations, and Expressly Adapted for the Yeomanry*, London, 1835
Arentschildt, Lt. Col. von, *Instructions for Officers and Non-Commissioned Officers of Cavalry, on Outpost Duty*, London, 1854
*The Army Reform Wanted – Maximum of Defence, with Minimum of Expence*, London, 1851
An Artillery Officer, *Considerations on Tactics*, London, 1853
*Considerations on Tactics ... Part II, Observations on Strategy*, London, 1854
Bedford, Maj. W. D., *Some Suggestions for the Cheap Defence of the Kingdom*, London, 1853.
Bismark, Count von, '*Lectures on the Tactics of Cavalry*', trans. by Maj. N. Ludlow Beamish, London, 1827
Boucher, John, *The Volunteer Rifleman, and the Rifle*, London, 1854
B[owles], W., *Thoughts on National Defence*, London, 1848
Boxer, Capt. E. M., *Treatise on Artillery, Prepared for the Use of the Practical Class, Royal Military Academy*, London, 1853
Brack, F. de, *Advance Posts of Light Cavalry*, trans. by Maj. P. J. Begbie, Madras, 1850
Burlton, Lt. Col. W., *A Few Brief Comments on Sir Charles Napier's Letter to Sir J. Hobhouse, 'On the Baggage of the Indian Army'*, London, 1849
Burton, Capt. J. Ryder, R.N., *A Letter Addressed to the Editor of the 'Morning Herald' on the National Defences*, London, 1848
Burton, Lt. Richard F., *A Complete System of Bayonet Exercise*, London, 1853
Bustin, W. R., *A Militia: Its Relation to the Regular Army*, London, 1847
Campbell, Lt. Col. James, *A British Army, as it Was, – Is, – and Ought to Be*, London, 1840
Chesney, Col. F. R., *Observations on the Re-construction of the Royal Regiment*

*of Artillery*, London, 1849

*Observations on the Past and Present State of Fire-arms ... with a Proposition for Reorganising the Royal Regiment of Artillery*, London, 1852

A Cidevant Cavalry Officer, *Army Reform: A Practical method of Reducing the Army Estimates a Million, Without Diminution of its Numerical Force*, London, 1833

A Civilian, *Hints for a Volunteer Coast Defence*, London, 1853

Connolly, T. W. J., *History of the Royal Sappers and Miners*, 2 vols., London, 1857

Craufurd, Robert, *Standing Orders as Given out and Enforced by the late Major-General Robert Craufurd, for the Use of the Light Division*, Dublin, 1844

Curling, Henry, *A Few Words in Recommendation of the Formation of Volunteer Rifle Corps*, London, 1850

David, *Letters to the Editor of 'The Times', in Answer to those of 'Emeritus'*, London, 1849

Delafield, Maj. Richard, *Report on the Art of War in Europe in 1854, 1855, and 1856*, Washington, 1860

de Ros, Lord, *A Selection of Field Movements for a Division of Cavalry*, London, 1846

*The Young Officer's Companion*, London, 1851

*Yeomanry Regulations*, London, 1853

Douglas, Gen. Sir Howard, *An Essay on the Principles and Construction of Military Bridges*, London, 1853

Douglas, Lt. Gen. Sir Howard, *A Treatise on Naval Gunnery*, 3rd edn, London, 1851

Downes, Alfred, *Manual of Information and Instruction of Candidates for Commissions*, London, 1852

Doyle, Lt. J. S., *The Military Catechism for the Use of Young Officers and Serjeants of Infantry*, London, 1825

Drummond, Henry, *A Letter to Captain Fitzroy on Rifle Corps*, London, 1852

Du Bourdieu, F. R. H., *Instructions for the Choice, Fortifying, Occupation, Defence and Attack of Military Positions*, London, 1830

Dupin, Charles, *View of the History and Actual State of the Military Force of Great Britain*, 2 vols., London, 1822

An Englishman and a Civilian, *The Invasion of England*, London, 1852

*Facts in Refutation of Some of the Misstatements Respecting the Board of Ordnance, in the ... Quarterly Review*, London, 1848

Faddy, Lt. Col., *Essay on the Defence and Military System of Great Britain, at Home and Abroad*, London, 1848

Favé, Capt., *The Emperor Napoleon's New System of Field Artillery*, trans. by Capt. W. H. Cox, London, 1854

Fenwick, Maj. Horace, *Essays on Field Fortification*, London, 1833

Fenwick, Capt. Horace, *The 'Field Exercise and Evolutions of the Army' Illustrated in a Series of Field Days*, London, 1853

Fergusson, James, *An Essay on a Proposed New System of Fortification: With Hints for its Application to our National Defences*, London, 1849

*The Peril of Portsmouth*, London, 1852

Fielding, Newton, *Lessons on Fortification*, London, 1853

Fitzclarence, Lt. Gen. Lord Frederick, *A Manual of Out-post Duties*, London, 1851

*Memoir on the Duty of Picquets*, London, 1843

*Memoranda by Lieut.-General Lord Frederick Fitzclarence, for the Use of*

*Young Officers Assembled in Poona during the Period of Exercise in December 1853, including Instructions Applicable to a Combination of the Three Arms*, London, 1854
*Suggestions for Brigade and Light Infantry Movements*, London, 1851
[Fletcher, Quartermaster-Sergeant John,] A Non-Commissioned Officer, *Advice to the British Soldier*, London, 1839
Flood, Capt. Warden, *A Sketch of the Military and Political State of Prussia*, Portsmouth, 1833
Fox, Lt. Col. Lane, *On the Improvement of the Rifle, as a Weapon for General Use*, London, 1858
Fyers, Capt., *Notes on the Defensive Resources of Great Britain*, 1852
Gardiner, Maj. Gen. Sir Robert, *Illustrations, in Questions and Answers, on the Numerical Deficiency, Want of Instruction, and Inefficient Equipment of the Artillery of the British Army*, [London], 1849
*Report on the Numerical Deficiency, Want of Instruction, and Inefficient Equipment of the Artillery of the British Army* [London], 1848
Gawler, Col. G., *The Essentials of Good Skirmishing*, London, 1852
Gilbert, Lt. J. A., *An Exposition of the First Principles of Grand Military Combinations and Movements*, London, 1825
Glover, R. Mortimer, *National Defence*, London, 1853
Gordon, Lt. Col. the Hon. A., *Remarks on National Defence, Volunteers, and Rifles*, London, 1853
Gore, Capt. John, *The Evolutions of a Field Battery*, Woolwich, 1846
Gore, Montague, *National Defences*, London, 1852
Greenwood, George, *Cavalry Sword Exercise*, [no imprint,] 1840
Grey, Third Earl, *Speech of the Rt. Hon. Earl Grey on the Second Reading of the Militia Bill*, London, 1852
Griffiths, Capt. F. A., *The Artillerist's Manual and Compendium*, Woolwich, 1839, 1840, 1847, 1854, 1856
Guthrie, G. J., *Commentaries on the Surgery of the War in Portugal, Spain, France, and the Netherlands*, London, 1853, 1855
Hamley, E. B., *The Operations of War*, Edinburgh, 1866
Hardbargain, Henry, *Hints to Subalterns of the British Army*, London, 1843
Harvey, Maj. H. B., *A Visit to the Camp of Beverloo*, London, 1852
Head, Sir Francis, *The Defenceless State of Great Britain*, London, 1850
*He Would not Neglect the Defence of the Country*, London, 1845
Hill, Frederick, *National Force, Economical Defence of the Country from Internal Tumult and Foreign Aggression*, London, 1848
*A History of the Sudden and Terrible Invasion of England by the French, in the Month of May 1852*, London, 1853
Hodge, Thomas, *Catechism of Fortification*, London, 1851
Hodge, William Barwick, 'On the Mortality Arising from Military Operations', *Journal of the Statistical Society of London*, XIX, 1856, 219
Horse Guards, *The Book of Aids, or Catechism in the System of Equitation, Practised at the Cavalry Riding Establishment*, London, 1831
*Cavalry Sword Exercise*, London, 1843
*Field Exercise and Evolutions of the Army*, London, 1824, 1833
*The Infantry Manual*, London, 1854
*Infantry Sword Exercise*, London, 1845

*Instructions for the Carbine Exercise; The Pistol Exercise; and the Lance Exercise*, London, 1844
*Instruction of Musketry*, London, 1855
*The King's Regulations and Orders for the Army*, London, 1837
*The Manual and Platoon Exercises*, London, 1834
*The Platoon Exercise, and Different Firings for the Ordinary or Rifle Musket*, London, 1854
*The Queen's Regulations and Orders for the Army*, London, 1844
*Regulations for Encampments*, London, 1853
*Regulations for the Instruction, Formations and Movements of the Cavalry*, London, 1833, 1844, 1851
Hough, Capt. William, *The Practice of Courts-Martial, and other Military Courts*, London, 1834
Humfrey, Lt. Col. I. H., *An Essay on the Modern System of Fortification*, London, 1838
Jackson, Robert, *A View of the Formation, Discipline and Economy of Armies*, London, 1845
Jacob, Col. John, *Rifle Practice*, London, 1858
Jacobi, G.-A., *Etat actuel de l'artillerie de campagne en Europe: artillerie de campagne anglaise*, trans. by Capt. Mazé, Paris, 1838
Jebb, Lt. Col. J., *A Flying Shot at Fergusson*, London, 1853
*Manual for the Militia*, London, 1853
*A Practical Treatise on the Attack of Military Posts, Villages, Intrenchments, etc.*, London, 1848
*A Practical Treatise on the Duties Required to be Performed by Officers and Soldiers of the Army at a Siege*, London, 1849
*Practical Treatise on Strengthening and Defending Outposts, Villages, Houses, Bridges, etc.*, London, 1848
Jervis, Lt. J.-W., *Manual of Field Operations*, London, 1852
Jervis, Capt. J.-W., *The Rifle Musket*, London, 1854
Jomini, A. H., *The Art of War*, trans. from *Précis de l'art de la guerre* by G. H. Mendell and W. P. Craighill, Philadelphia, 1875
Jones, Maj. Inigo (trans. and ed.), *The Three Arms, or, Divisional Tactics, of Decker*, London, 1848
Kaye, Edward, *Memoir of Some Proposed Improvements in Modern Fortification*, Calcutta, 1836
Ker, Capt., *Proposed System of Out-post Exercise for Regiments of Cavalry and Infantry*, London, 1827
Kimber, Thomas, *The Construction of the Modern System*, London, 1852
*Construction of Vauban's First System*, London, 1851
Kinglake, A. W., *The Invasion of the Crimea*, 8 vols., Edinburgh, 1863–88
Kinloch, John, *Proposal for the Defence of the Country*, Edinburgh, 1852
Knox, Capt. Charles, *The Defensive Position of England*, London, 1852
Landmann, Isaac, *The Principles of Fortification*, London, 1831
Le Couteur, Col. John, *The Rifle*, London, 1855
Lefroy, Capt. J. H. (ed.), *A Handbook for Field Service*, London, 1854
Lendy, A. F., *The Principles of War*, London, 1853
Leslie, Maj. C., *Instructions for the Application of Light Drill to Skirmishing in the Field*, Dublin, 1831
Long Range, *The Rifle*, London, 1852

Lopez Suasso Diaz de Fonseca, Baron Don Antonio, *The Theory of Infantry Movements*, 2 vols., London, 1846
Luard, Lt. Col. John, *A History of the Dress of the British Soldier*, London, 1852
Lyster, Capt. F. Torrens, *Battalion Drill Designed for the Use of ... the Militia*, London, 1853
*Manual of Information for Regimental Officers*, London, 1853
Macaulay, Capt. J. S., *A Treatise on Field Fortification*, London, 1847
McClellan, George B., *Report of the Secretary of War, Communicating the Report of Captain George B. McClellan, ... One of the Officers Sent to the Seat of War in Europe in 1855 and 1856*, Washington, 1857
MacDougall, Patrick L., *The Theory of War*, London, 1856
Macfarlane, Charles, *The Camp of 1853*, London, 1853
Macgregor, Capt. Robert S., *A Plan for Raising a Defensive Force*, Edinburgh, 1849
Macintosh, John, *New Strategies in Warfare*, London, 1853
[McMullen, J.,] A late staff sergeant of the 13th Light Infantry, *Camp and Barrack-room*, London, 1846
McMurdo, Maj. Montagu, *Sir Charles Napier's Indian Baggage-Corps: Reply to Lieut. Col. Burlton's Attack*, London, 1850
Macnamara, Maj. Ulysses, *The British Army*, London, [1839]
Magrath, Lt. R. N., *An Historical Sketch of the Progress of the Art of War*, Dublin, 1838
Maitland, Lt. Col. Fred. T., *Theoretical Instruction in Ball Practice with Musket and Minié*, London, 1854
Malins, William, *A Plan for Additional National Defences*, London, 1848
*A Manual of Position Drill*, London, 1850
Martin, Lt., *Guide to the Military Examination*, London, 1851
Maurice [de Sellon], Baron P.-E., *On National Defence in England*, London, 1852
Mayne, Capt. William C., *Military Reconnaissance*, Exeter, 1837
Metius Curtius, *Essay on the Constitution of the British Artillery* [London], 1848
Miles, *War! How Best Met, or Prevented*, Edinburgh, 1846
A Military Tutor, *The Pattern Military Officer*, London, 1855
Mitchell, Lt. Col. John, *Thoughts on Tactics*, London, 1838
Mitchell, Lt. T. H., *A System of Light Drill*, London, 1845
Mitchell, Maj. T. L., *Series of Figures Shewing all the Motions in the Manual and Platoon Exercises, and Different Firings, According to His Majesty's Regulations*, London, 1828 (reprint, Ottawa, 1965)
Mordecai, Alfred, *Military Commission to Europe, in 1855 and 1856: Report of Major Alfred Mordecai of the Ordnance Department*, Washington, 1860
Mountsteven, Capt. W. J. B., *Observations on the Infantry Manoeuvres of the British Army*, London, 1842
Napier, Lt. Gen. Sir C. J., *Defects, Civil and Military, of the Indian Government*, London, 1853
*A Letter on the Defence of England, by Corps of Volunteers and Militia*, London, 1852
*A Letter to the Rt. Hon. Sir J. Hobhouse on the Baggage of the Indian Army*, London, 1849
*Lights and Shades of Military Life*, 2 vols., London, 1840
*Remarks on Military Law and the Punishment of Flogging*, London, 1837

[Napier, Elers,] *Past and Future Emigration; or the Book of the Cape*, London, 1849
Napier, Maj. Gen. Sir William, *Six Letters, in Vindication of the British Army, Exposing the Calumnies of the Liverpool Financial Reform Association*, London, 1849
*The National Defences*, London, 1845
Ney, Marshal, *Military Studies by Marshal Ney*, trans. by G. H. Caunter, introduced by Maj. A. James, London, 1833
Nolan, Capt. L. E., *Cavalry: Its History and Tactics*, London, 1853
*The Training of Cavalry Remount Horses*, London, 1852
Oakes, A. F., *Instructions and Regulations for Artillery*, Madras, 1847
Oakes, Capt. A. F., *The Young Artillery Officer's Assistant*, Madras, 1847
*Observations upon the Peace Establishments of the Army*, London, 1822
O'Doodle, Sir Phelim, *The Subaltern's Check-book*, London, 1848
An Officer, *A Few Words on the Defence of the Country from Invasion*, London, 1846
*Observations on the Army*, London, 1825
An Officer of the Corps, *The Royal Artillery: A Few Remarks on its Deficiency and Inefficiency*, London, 1849
*On the Deteriorated Condition of our Saddle-horses; and the Causes and the Remedy: The State of our Cavalry, and the Imperfect System under which this Force and that of our Army Generally is Administered*, London, 1853
Ordnance, Board of, *A Course of Elementary Fortification*, by C. W. Pasley, 2 vols., London, 1822
*Exercise of the Cylindrical Pontoon, According to the Practice of the Royal Engineer Establishment*, Chatham, 1852
*Instructions and Regulations for the Service and Management of Heavy Ordnance for the Royal Regiment of Artillery*, Woolwich, 1849
*Orders and Regulations for the Guidance of the Corps of the Royal Engineers and Royal Sappers and Miners*, London, 1831
*Regulations and Instructions for the Royal Artillery on Matters of Finance, and Points of Discipline connected therewith*, London, 1850
*Standing Orders and Regulations for the Royal Regiment of Artillery at Home and Abroad*, London, 1847
[Ordnance, Board of: Royal Engineer Establishment,] *Of the Attack of Fortified Places* [Chatham?, 1822?]
Ordnance, Board of: Royal Engineer Establishment, *Detail of Siege Operations Proposed for the Garrison of Chatham, in Concert with the Royal Engineers* ... Chatham, 15 June 1837, 24 June 1839, 10 May 1841, 29 August 1843
*The Exercise of Military Bridges of Casks*, Chatham, 1822
*Exercise of the New Decked Pontoons, or Double Canoes, invented by Lieut. Colonel Pasley, R.E.*, Chatham, 1823
*Method of Loading and Throwing Hand Grenades*, Chatham, 1821
*Practical Rules for making Telegraphic Signals*, by Lt. Col. C. W. Pasley, Chatham, 1822
*Rules for Escalading Works of Fortification not having Palisaded Covered Ways*, Chatham, 1822
*A Simple Practical Treatise on Field Fortification for the Use of the Regimental Schools of the Royal Sappers and Miners*, Chatham, 1823
Para Bellum, *Brief Suggestions on the Subject of War and Invasion*, London, 1852

Pasley, Maj. Gen. C. W., *Rules for Conducting the Practical Operations of a Siege*, 2 vols., London, 1841–2
Percy, Lt. Col. the Hon. H., *Explanations of the Manoeuvres of a Brigade of Infantry*, London, 1852
Peters, J. G., *A Treatise on Equitation*, London, 1835
Prior, L. Murray, *A Concise Manual of Field Exercises and Evolutions for Volunteer Rifle Corps*, London, 1852
Pro Patria, *Suggestions for the formation of a Corps of Volunteer Cavalry* [London], 1854
Raines, Lt. Col. J. R., *A Guide to the Organisation and Training of the Militia, Yeomanry, Infantry, and Volunteer Corps*, London, [1852]
Ranelagh, Viscount, *Observations on the Present State of our National Defences*, London, 1845
A Regimental Officer, *A Few Remarks about the British Army*, London, 1857
*Remarks on the Education of the Royal Artillery* [possibly by Capt. F. Eardley-Wilmot], Edinburgh, 1848
A Retired Artillery Officer, *The Defence of our Mercantile Seaports*, London, 1852
Robbins, Lt., *The Cavalry Catechism*, London, 1851
Rolt, Col. John, *Moral Command*, London, 1842
Scoffern, John, *Projectile Weapons of War*, London, 1852
Shirley, Lt. Col. Arthur, *Remarks on the Transport of Cavalry and Artillery*, London, 1854
Short, Capt. Charles W., *A Treatise on the Disposition and Duties of Out-Posts*, London, 1827
Short, Lt. Col. Charles W., *A Treatise on Patroling*, London, 1831
Short, Charles W., *Vade Mecum of the Subaltern and Non-Commissioned Officer on Outpost Duty*, London, 1854
Simmons, Capt, T. F., *Ideas as to the Effect of Heavy Ordnance*, London, 1837
[Simmons, Lt. T. F.,] *Remarks on the Promotion of the Officers of the Corps of Artillery in the British Service, and on the Application of that Arm in the Field*, London, 1819
Sinnot, Lt. John, *A Manual of Light Infantry and Other Duties*, London, 1849
*A Military Catechism*, Portsmouth, 1841
Smith, John, *A System to Organize and Establish Military Discipline, in a Patriotic or Defensive Army*, London, 1845
Snell, William Henry, *Company and Battalion Drill*, London, 1849 and 1850
Sparks, Maj. J. P., *Details of Light Infantry Drill, with Some Instruction for Patrols*, London, [1845]
Sparks, Lt. Col. J. P., *Tracts on Infantry Tactics*, No. 1, London, 1854
Spearman, Capt. J. Morton, *The British Gunner*, London, 1850.
*Notes on Military Education*, London, 1853
Stocqueler, J. H., *The British Officer*, London, 1851
*Familiar History of the British Army*, London, 1871
*The Military Encyclopaedia*, London, 1853
*A Personal History of the Horse Guards from 1750 to 1872*, London, 1873
Straith, Maj. Hector, *Introductory Essay to the Study of Fortification*, London, 1852
*Treatise on Fortification and Artillery*, 2 vols., London, 1852
*Suggestions for the Defence of Great Britain Without Increased Expenditure*, London, 1851

Thackeray, Capt., *The Soldier's Manual of Rifle Firing at Various Distances*, London, 1854
*Three Lectures Delivered at the Guildhall, Bath, on the Practice of Rifle Firing at Various Distances*, London, 1853
Torrens, Col. A. W., *Notes on the French Infantry*, London, 1852
[Torrens, A. W.,] A Field Officer, *Six Familiar Lectures, for the Use of Young Military Officers*, London, 1851
Troubridge, Maj. [T.] St Vincent, *Principles of the Minor Operations of War: Translated from the French of Lallemand*, London, 1852
Troubridge, Capt. T. St Vincent, *Tabular Arrangement of Battalion Drill*, London, [1851]
Turner, Cpl Maj. Robert, *A Practical Guide to the Breaking and Training of the Young Horse*, London, 1851
Walshe, Capt. Anthony, *A Catechism and Handbook on Regimental Standing Orders*, London, 1852
[Wathen, A.?] *Remarks on the Proposed Regulations for the Instruction, Formations and Movements of the Cavalry*, Dorking and London, 1832
Weale, John, *Letter to the Rt. Hon. Lord John Russell, on the Defence of the Country*, London, 1847
Wilkinson, Henry, *Engines of War*, London, 1841
*Observations on Swords*, London, [1844]
Yates, Edward, *Elementary Treatise on Strategy*, London, 1852
*Elementary Treatise on Tactics and on Certain Parts of Strategy*, London, 1853
Yule, Lt. Henry, *Fortification for Officers of the Army and Students of Military History*, Edinburgh, 1854

### V. Diaries, memoirs, etc.

Alexander, Sir James E., *L'Acadie*, 2 vols., London, 1849
*Narrative of a Voyage of Observation among the Colonies of Western Africa, in the Flag-ship Thalia; and of a Campaign in Kaffir-land, on the Staff of the Commander-in-Chief, in 1835*, 2 vols., London, 1837
*Passages in the Life of a Soldier*, 2 vols., London, 1857
Anderson, Lt. Col. Joseph, *Recollections of a Peninsular Veteran*, London, 1913
Anglesey, Marquess of (ed.), *Little Hodge*, London, 1971
*Sergeant Pearman's Memoirs*, London, 1968
Astley, Sir John Dugdale, *Fifty Years of My Life*, 2 vols., London, 1894
Baird, J. G. A. (ed.), *Private Letters of the Marquess of Dalhousie*, Edinburgh, 1910
[Bayley, J. A.,] *Reminiscences of School and Army Life, 1839 to 1859*, privately printed, 1875
Benson, A. C., and Esher, Viscount, *The Letters of Queen Victoria 1837–1861*, 3 vols., London, 1907
Bisset, Maj-Gen. [J.], *Sport and War*, London, 1875
Brialmont, M., and Gleig, the Rev. G. R., *History of the Life of Arthur, Duke of Wellington*, 4 vols., London, 1860
Bruce, H. A. (ed.), *Life of General Sir William Napier, K.C.B.*, 2 vols., London, 1864
Bunbury, Thomas, *Reminiscences of a Veteran*, 3 vols., London, 1861

Cathcart, Sir George, *Correspondence of Lieut-General the Hon. Sir George Cathcart, K.C.B., relative to his Military Operations in Kaffraria*, London, 1856
Chesney, L. and J., *The Life of the late General F. R. Chesney*, London, 1893
Combermere, Mary, Viscountess, *Memoirs and Correspondence of Field-Marshal Viscount Combermere*, 2 vols., London, 1866
Cumming, Lt. James S., *A Six Years' Diary*, London, 1847
de Ainslie, Gen. [C.], *Life as I have Found it*, Edinburgh, 1883
Douglas, Sir George, and Ramsay, Sir George Dalhousie (ed.), *The Panmure Papers*, 2 vols., London, 1908
Douglas, Pte. William, *Soldiering in Sunshine and Storm*, Edinburgh, 1865
Eardley-Wilmot, Mrs F. M., *Memorials of Fredk. M. Eardley-Wilmot, Maj.-Gen., R.A., and F.R.S.*, London, 1879
Ewart, Lt. Gen. J. A., *The Story of a Soldier's Life*, 2 vols., London, 1881
Fergusson, William, *Notes and Recollections of a Professional Life*, edited by James Fergusson, London, 1846
Ferrar, M. L. (ed.), *Diary of Colour-Sergeant George Calladine, 19th Foot, 1793–1837*, London, 1922
Fullom, S. W., *The Life of General Sir Howard Douglas*, London, 1863
Gleig, G. R., *Personal Reminiscences of the 1st Duke of Wellington*, Edinburgh, 1904
Gordon-Brown, A. (ed.), *The Narrative of Private Buck Adams*, Cape Town, 1941
Hall, J. H. W., *Scenes in a Soldier's Life*, Montreal, 1848
Henderson, R., *The Soldier of Three Queens*, 2 vols., London, 1866
Higginson, Gen. Sir George, *Seventy-one years of a Guardsman's Life*, London, 1916
Humbley, Capt. W. W. W., *Journal of a Cavalry Officer*, London, 1854
Jagow, Kurt (ed.), *Letters of the Prince Consort 1831–1861*, London, 1938
Kealy, P. H., *General Sir Charles William Pasley* [Chatham, 1932]
King, Capt. W. R., *Campaigning in Kaffirland*, London, 1853
Knollys, Col. Henry (ed.), *Life of General Sir Hope Grant*, 2 vols., Edinburgh, 1894
Leach, Lt. Col. J., *Rambles along the Styx*, London, 1847
Lefroy, Lady (ed.), *Autobiography of General Sir John Henry Lefroy, C.B., K.C.M.G., F.R.S.*, privately printed, 1895
Lysons, Gen. Sir Daniel, *Early Reminiscences*, London, 1896
Mackenzie, Lt. Gen. Colin, *Storms and Sunshine of a Soldier's Life*, Edinburgh, 1884
Marshman, John C., *Memoirs of Major-General Sir Henry Havelock, K.C.B.*, London, 1876
Martin, Theodore, *The Life of His Royal Highness the Prince Consort*, London, 1875–6
Martineau, John, *The Life of Henry Pelham, 5th Duke of Newcastle*, London, 1908
Mitra, S. M., *The Life and Letters of Sir John Hall*, London, 1911
Mountain, Mrs. A. S. H. (ed.), *Memoirs and Letters of the late Colonel Armine S. H. Mountain, C.B.*, London, 1858
Munro, Surgeon-Gen. William, *Records of Service and Campaigning in Many Lands*, 2 vols., London, 1887
Napier, Lt. Col. E. Elers, *Excursions in Southern Africa*, 2 vols., London, 1849

Napier, Col. Elers, *The Linesman*, 3 vols., London, 1856
Napier, Lt. Gen. Sir W. F. P., *The Life and Opinions of General Sir Charles James Napier, G.C.B.*, 4 vols., London, 1857
Neill, J. M. B., *Recollections of Four Years' Service in the East with H.M. Fortieth Regiment*, London, 1845
Orlich, Capt. Leopold von, *Travels in India, including Scinde and the Punjab*, London, 1845
Patterson, Maj. John, *Camp and Quarters*, 2 vols., London, 1840
Rait, Robert S., *The Life and Campaigns of Hugh, 1st Viscount Gough, Field Marshal*, 2 vols., London, 1903
Ramsay, Lt. Col. B. D. W., *Rough Recollections of Military Service and Society*, 2 vols., Edinburgh, 1882
[Reid, Sgt. David,] *Memorials of the Life of a Soldier*, Edinburgh, 1864
Robertson, Col. James P., *Personal Adventures and Anecdotes of an Old Officer*, London, 1906
[Ryder, John,] A Private Soldier, *Four Years' Service in India*, Leicester, 1853
Sanderson, Charles, R. (ed.), *The Arthur Papers*, 3 vols., Toronto, 1957–9
Shadwell, Lt. Gen., *The Life of Colin Campbell, Lord Clyde*, 2 vols., Edinburgh, 1881
Sidney, Rev. Edwin, *The Life of Lord Hill*, London, 1845
[Smith, Dr.,] *The Christian Soldier: Being a Memorial of the late Lieut.-Colonel Fordyce*, London, [? 1855]
Smith, G. C. Moore, *The Autobiography of Lieutenant-General Sir Harry Smith*, London, 1901
*The Life of John Colborne, Field Marshal Lord Seaton*, London, 1903
[Somerville, A.,] *The Autobiography of a Working Man*, London, 1848
Stanhope, Philip Henry, Fifth Earl, *Notes of Conversations with the Duke of Wellington 1831–1851*, London, 1938
Stuart, Col. W. K., *Reminiscences of a Soldier*, 2 vols., London, 1874
Swinson, Arthur, and Scott, Donald (ed.), *The Memoirs of Private Waterfield*, London, 1968
Taylor, Sgt. Maj. [William], *Life in the Ranks*, London, 1843
Thackwell, Edward J., *Narrative of the Second Seikh War*, London, 1851
Tulloch, Maj. Gen. Sir A. B., *Recollections of Forty Years' Service*, Edinburgh, 1903
Verner, Col. Willoughby, *The Military Life of H.R.H. George, Duke of Cambridge*, 2 vols., London, 1905
*The War in India*, despatches of Hardinge, Gough and Smith, London, 1846
Ward, S. G. P. (ed.), *The Hawley Letters*, London, 1970
Watts, Alaric Alfred, *Alaric Watts: A Narrative of His Life*, 2 vols., London, 1884
Wellington, Second Duke of, *Despatches, Correspondence and Memoranda of Field Marshal Arthur, Duke of Wellington, K.G.*, 8 vols., London, 1867–80
Whittingham, Maj. Gen. Ferdinand (ed.), *A Memoir of The Service of Lieut.-General Sir Samuel Ford Whittingham K.C.B., K.C.H., G.C.F.*, London, 1868
Wolseley, Field-Marshal Viscount, *The Story of a Soldier's Life*, 2 vols., London, 1903
Wood, Evelyn, *The Crimea in 1854 and 1894*, London, 1895
Wrottesley, Lt. Col. the Hon. George, *Life and Correspondence of Field Marshal Sir John Burgoyne, Bart.*, 2 vols., London, 1873

*The Military Opinions of General Sir John Fox Burgoyne, Bart., G.C.B.*, London, 1859
Wylly, Col. H. C., *The Military Memoirs of Lieut. General Sir Joseph Thackwell*, London, 1908

### VI. Secondary works

Almack, Edward, *The History of the Second Dragoons 'Royal Scots Greys'*, London, 1908
Anglesey, Marquess of, *A History of the British Cavalry 1816 to 1919*, 2 vols., London, 1973–5
*One Leg*, London, 1961
Arthur, Capt. Sir George, *The Story of the Household Cavalry*, vol. II, London, 1909.
Atkins, J. B., *The Life of Sir William Howard Russell*, London, 1911.
Atkinson, C. T., *The Dorsetshire Regiment*, 2 vols., Oxford, 1947
*History of the Royal Dragoons*, Glasgow, 1934
*Regimental History, The Royal Hampshire Regiment*, vol. I, Glasgow, 1950
*The South Wales Borderers, 24th Foot, 1689–1937*, Cambridge, 1937
Austin, M., *The Army in Australia 1840–50*, Canberra, 1979
Barrett, C. R. B., *The 85th King's Light Infantry*, London, 1913
*History of the XIII Hussars*, vol. I, Edinburgh, 1911
Bartlett, C. J., *Great Britain and Sea Power 1815–1853*, Oxford, 1963
Bartlett, H. Moyse, *Louis Edward Nolan and his Influence on the British Cavalry*, London, 1971
Bayley, C. C., *Mercenaries for the Crimea*, Montreal, 1977
Baynes, John, *Morale*, London, 1967
Blackmore, Howard L., *British Military Firearms 1650–1850*, London, 1961
Blanco, Richard L., 'Reform and Wellington's Post Waterloo Army, 1815–1854', *Military Affairs*, XXIX, 1965, 123–31
Bolitho, Hector, *The Galloping Third*, London, 1963
Bond, Brian, *The Victorian Army and the Staff College, 1854–1914*, London, 1972
Callwell, Sir Charles, and Headlam, Sir John, *The History of the Royal Artillery from the Indian Mutiny to the Great War*, 2 vols., Woolwich, 1931–7
Cantlie, Sir Neil, *A History of the Army Medical Services*, 2 vols., Edinburgh, 1974
Cary, A. D. L., and McCance, Stouppe, *Regimental Records of the Royal Welch Fusiliers*, 2 vols., London, 1923
Caulfield, J. E., *One Hundred Years' History of the 2nd Batt. West India Regiment*, London, 1899
Chaplin, H. D., *The 97th or Earl of Ulster's Regiment 1824–1881*, Maidstone, 1973
Cowper, L. I. (ed.), *The King's Own*, 2 vols., Oxford, 1939
Crew, Graeme, *The Royal Army Service Corps*, London, 1970
Curtiss, John Shelton, *The Russian Army under Nicholas I, 1825–1855*, Durham, N.C., 1965
Dalbiac, P.H., *History of the 45th: 1st Nottinghamshire Regiment (Sherwood Foresters)*, London, 1902
Daniell, David Scott, *Cap of Honour*, London, 1951
Davidson, H., *History and Services of the 78th Highlanders (Ross-shire Buffs) 1793–1881*, Edinburgh, 1901

Day, E. C., and Adams, R. M., *Through to 1970*, London, 1970
Dunn-Pattison, R. P., *The History of the 91st Argyllshire Highlanders*, Edinburgh, 1910
Dupree, Louis, 'Afghan and British Military Tactics in the First Anglo-Afghan War (1838–1842)', *Army Quarterly*, CVII, 1977, 214–21
Forbes, A., *A History of the Army Ordnance Services*, 3 vols., London, 1929
Fortescue, J. W., 'The Army', in G. M. Young (ed.), *Early Victorian England 1830–1865*, 2 vols., London, 1934
*The British Army 1783–1802*, London, 1905
*A History of the British Army*, vols. XI–XIII, London, 1923–30
*A History of the 17th Lancers*, London, 1895
*The Last Post*, Edinburgh, 1934
*The Royal Army Service Corps: A History of Transport and Supply in the British Army*, vol. I, Cambridge, 1930
*Six British Soldiers*, London, 1928
Fuller, J. F. C., *Sir John Moore's System of Training*, London, 1924
Fyler, Col., *History of the 50th or (The Queen's Own) Regiment*, London, 1895
Gardyne, C. Greenhill, *The Life of a Regiment*, 2 vols., Edinburgh, 1903
Glover, Richard, *Peninsular Preparation*, Cambridge, 1963
Godwin-Austen, A. R., *The Staff and the Staff College*, London, 1927
Gooch, Brison D., *The New Bonapartist Generals in the Crimean War*, The Hague, 1959
Graham, Henry, *History of the 16th, the Queen's Light Dragoons (Lancers), 1759 to 1912*, Devizes, 1912
Gretton, G. le M., *The Campaigns and History of the Royal Irish Regiment, from 1684 to 1902*, Edinburgh, 1911
Griffith, P. G., 'Military Thought in the French Army 1815–1851', Oxford University D.Phil. Thesis, 1975
Groves, Percy, *Historical Records of the 7th or Royal Regiment of Fusiliers*, Guernsey, 1903
Guggisberg, F. G., *'The Shop', The Story of the Royal Military Academy*, London, 1900
Hamilton, C. I., 'The Lancaster Gun', unpublished paper
'The Royal Navy, Seapower, and the Screw Ship of the Line, 1845–1860', Cambridge University Ph.D. thesis, 1974
Hardinge, Charles, Viscount, *Rulers of India: Viscount Hardinge*, Oxford, 1891
Harris, John, *The Gallant Six Hundred*, London, 1973
Hay, George Jackson, *An Epitomized History of the Militia (the 'Constitutional Force')*, London, 1906
Hibbert, Christopher, *The Destruction of Lord Raglan*, London, 1961
Hime, H. W. L., *History of the Royal Regiment of Artillery 1815–1853*, London, 1908
Hogg, O. F. G., *Artillery: Its Origin, Heyday and Decline*, London, 1970
*The Royal Arsenal. Its Background, Origin and Subsequent History*, 2 vols., London, 1963
Houlding, J. A., *Fit for Service: The Training of the British Army, 1715–1795*, Oxford, 1981
Howard, Michael, *The Franco-Prussian War*, London, 1961
*Studies in War and Peace*, London, 1970
Hughes, B. P., *The Bengal Horse Artillery 1800–1861*, London, 1971
*British Smooth-Bore Artillery*, London, 1969

*Firepower*, London, 1974
Irvine, Dallas D., 'The Origin of Capital Staffs', *Journal of Modern History*, x, 1938, no. 2, 161
Irwin, D. Hastings, *War Medals and Decorations*, London, 1890
Jocelyn, J. R. J., *The History of the Royal Artillery (Crimean Period)*, London, 1911
*The History of the Royal and Indian Artillery in the Mutiny of 1857*, London, 1915
Johnston, S. H. F., *The History of the Cameronians (Scottish Rifles)*, vol. I, Aldershot, 1957
*Journal of the Arms and Armour Society*
*Journal of the Society for Army Historical Research*
Laurie, G. B., *History of the Royal Irish Rifles*, London, 1914
Leask, J. C., and McCance, H. M., *The Regimental Records of the Royal Scots*, Dublin, 1915
Lehmann, Joseph H., *Remember You Are an Englishman: A Biography of Sir Harry Smith 1787–1860*, London, 1977
Liddell, R. S., *The Memoirs of the Tenth Royal Hussars*, London, 1981
Linklater, Eric and Andro, *The Black Watch*, London, 1977
Lomax, D. A. N., *History of the Services of the 41st (the Welch) Regiment*, Devonport, 1899
Lovell-Knight, A. Vaughan, *The History of the Office of the Provost Marshal and the Corps of Military Police*, Aldershot, 1943
Luvaas, Jay, *The Education of an Army*, London, 1965
Massé, C. H., *The Predecessors of the Royal Army Service Corps*, Aldershot, 1948
Maurice-Jones, K. W., *The History of Coast Artillery in the British Army*, London, 1959
Moorsom, W. S., *Historical Record of the 52nd Regiment (Oxfordshire Light Infantry)*, London, 1860
Morrell, W. P., *British Colonial Policy in the Age of Peel and Russell*, Oxford, 1930
*British Colonial Policy in the Mid-Victorian Age*, Oxford, 1969
Moyer, R. A., 'The Mfengu Self-defence, Expansion and the Frontier Wars', unpublished paper given at conference on the history of the Transkei and Ciskei
Murray, Robert H., *History of VIII King's Royal Irish Hussars 1693–1927*, Cambridge, 1928
Oatts, L. B., *I Serve*, Norwich, 1966
*Proud Heritage*, vols. I and II, Edinburgh, 1952–9
Oman, C. W. C., *Wellington's Army 1809–1814*, London, 1912
Pollock, J. C., *Way to Glory*, London, 1957
Pomeroy, Ralph Legge, *The Story of a Regiment of Horse*, vol. I, Edinburgh, 1924
Porter, Whitworth, *History of the Corps of Royal Engineers*, 2 vols., London, 1889
Pycroft, George, *The Origin of the Volunteer Movement of 1852 and History of the First Volunteer Rifle Corps*, London, 1881
Raikes, G. A., *Historical Records of the First Regiment of Militia; or, 3rd West York Light Infantry*, London, 1876
Roads, C. H., *The British Soldier's Firearm 1850–1864*, London, 1964
'The History of the Introduction of the Percussion Breech-Loading Rifle into British Military Service 1850–70', Cambridge University Ph.D. thesis, 1961
Robson, Brian, *Swords of the British Army*, London, 1975

Rogerson, W., *Historical Records of the 53rd (Shropshire) Regiment*, London, 1890

Rosa, Joseph G., *Colonel Colt, London*, London, 1976

Ross-of-Blandensburg, Lt. Col., *A History of the Coldstream Guards, from 1815 to 1895*, London, 1896

Rothenberg, Gunther E,. *The Art of Warfare in the Age of Napoleon*, London, 1977

*Royal Inniskilling Fusiliers December 1688 to July 1914*, London, 1928

Samut-Tagliaferro, A., *History of the Royal Malta Artillery*, vol. I, Malta, 1976

Scott, J. B., 'The British Soldier on the Eastern Cape Frontier 1800–1850', University of Port Elizabeth, Ph.D. thesis, 1973

Seaton, Albert, *The Crimean War: A Russian Chronicle*, London, 1977

Shaw, A. G. L. (ed.), *Great Britain and the Colonies 1815–1865*, London, 1970

Sheppard, E. W., *The Ninth Queen's Royal Lancers 1715–1936*, Aldershot, 1939

Showalter, Denis E., *Railroads and Rifles: Soldiers, Technology and the Unification of Germany*, Hamden, Conn., 1975

Skelley, Alan Ramsay, *The Victorian Army at Home*, London and Montreal, 1977

Smith, Sir Frederick, *A History of the Royal Army Veterinary Corps 1796–1919*, London, 1927

Smyth, B., *History of the XX Regiment 1688–1888*, London, 1889

Spiers, Edward M., *The Army and Society 1815–1914*, London, 1980

Stacey, C. P., *Canada and the British Army, 1846–1871*, London, 1936

Stewart, P. F., *The History of the XII Royal Lancers*, London, 1950

Strachan, Hew, 'The Early Victorian Army and the Nineteenth-century Revolution in Government, *English Historical Review*, XCV, 1980, 782–809

'Soldiers, Strategy and Sebastopol', *Historical Journal*, XXI, 1978, 303–25

and Michael Barthorp, 'The Battle of Meanee', *Journal of the Society for Army Historical Research*, LV, 1977, 127–34

Sweetman, John, 'Military Transport in the Crimean War, 1854–1856', *English Historical Review*, LXXXVIII, 1973, 81

Tancred, George, *Historical Record of Medals and Honorary Distinctions*, London, 1891

*War Medal Record*, 2 vols., London, 1896–8

Thomas, Donald, *Charge! Hurrah! Hurrah!*, London, 1974

Tylden, G., *Horses and Saddlery*, London, 1965

Vale, W. L., *History of the South Staffordshire Regiment*, Aldershot, 1969

Wake, Joan, *The Brudenells of Deene*, London, 1953

Ward, B. R., *The School of Military Engineering, 1812–1909*, Chatham, 1909

Ward, S. G. P., *Faithful*, Edinburgh [1963]

*Wellington's Headquarters*, Oxford, 1957

Webb, E. A. H., *A History of the Services of the 17th (the Leicestershire) Regiment*, London, 1911

Whinyates, F. A., *From Coruña to Sebastopol*, London, 1884

Whyte, Frederic, and Atteridge, A. Hilliard, *A History of the Queen's Bays*, London, 1930

Williams, Godfrey T. R., *The Historical Records of the Eleventh Hussars*, London, 1908

Woodham-Smith, Cecil, *The Reason Why*, Harmondsworth, 1958

Wylly, H. C., *XVth (the King's Hussars) 1759 to 1913*, London, 1914

*The Loyal North Lancashire Regiment*, vol. I, London, 1933

# Index